MENTAL FUNCTIONING
by
Richard Stevko

The Graven Image
Publishing

23 Stafford Road
Hampden, MA
01036

ISBN 978-1-304-11392-4

Also by Richard Stevko

ABSURDITIES -- b&w photos with annotations
ABSURDITIES II -- color photos with annotations
CONTRASTS -- postmodern poetry with guide
THE GHOST TREE - T'ang style poetry
in collaboration with painter, Tamara Stevko
WHERE THOUGHTS COME FROM -- modern poetry
BEYOND A PERSIMMON -- metered poetry

ACKNOWLEDGMENT

The author gratefully recognizes the efforts of Lorraine and Tamara, who rigorously proofread this manuscript, seeking typos as well as unclear descriptions. Their accomplishment is particularly valuable in that they, at the same time, absorbed the information and participated in discussions on the topics.

Tamara, particularly made available the skills of her design company, *Tamarartistry*, to make ready the diagrams and charts on complex points, rending invaluable assistance in designing clarification of muddles.

CONTENTS

PREFACE
FIVE DECADES PLUS HALF-A

Fifty five years ago , as I completed work on my baccalaureate degree, I was confronted with the decision to apply for postgraduate work in medical school or in the Department of Philosophy. Reducing the conflict to its basic components, what I really wanted to study was: how the mind works. I chose Medicine as having more reliable answers. I was wrong in that it did not and still doesn't; but was right in that it is a more empiric approach. So - from the start, I was more of an Aristotelian; but my early religious training was more Platonic, by way of Aquinas. Many revisions of Plato followed - each yielding different answers (sic). The original question remains, neither Medicine nor Philosophy has yielded a satisfactory answer, and the following work is offered to help subsequent workers avoid the pitfalls into which I fell; and from which I learned the most.

GOALS

The goal of this work is to enumerate the mental functions (the Mind). Most of those functions are usually considered abstract, but the anatomic substrate upon which they depend (the brain) is concrete. The current definitions are summarized:

CONCRETE - a tangible object that has spatial location and causal powers
ABSTRACT - an idea or concept that lacks spatial location

	noun	adjective	verb	Piaget	Plato	problem
abstract (general)	an object lacking causal powers or spatial location	a quality or attribute apart from application to a particular object or instance	separated from embodiment, ideal	mental process (an idea or concept)	Ideal	how do we know about them?how can they affect our sensory experience?
concrete (particular)	physical objects are tokens of a particular type of thing. The "type" that it is a part of itself is an abstract object.	realities (physical things, sensations, or emotions) characterized by immediate experience	to make actual or real	tangible objects	Real	general/ specific, problem of universals

A secondary goal is to attempt to unify the functions with their substrates.
A tertiary goal is to determine which functions of the mind are authentic, and not factitious or contrived as riders.

INTRODUCTION

Today, every person who has lived in a literate country and has attended public school, has at least a passable amount of knowledge about how the body is constructed (anatomy) and how it works (physiology). A sizable percentage of those people go on to study the basics of health care, and are required to learn even more than a modicum
of these basic life sciences; sometimes including how it goes wrong (pathology).

These disciplines are so widely accepted, that the information is almost axiomatic, i.e., taken for granted. That was not always true. The bits of information (concrete) were learned from their experience with living beings (especially in hunting, butchering livestock, war) were often transmitted orally, accreting to other related bits of information as a discipline (abstract) and became historic knowledge only after writing developed. Even that was not a smooth process. One of the earliest writers, Homer, perpetuated some misinformation. His description of thoracic injuries and their effects are as useful as some autopsy reports, while other wounds are healed magically. His attribution of emotions (the main theme of the *ILIAD*), is replete with metaphors which have become cultural references: *fire in the belly, swelling chest, iron in the heart*. At the time, many citizens considered these to be literal physiology.

It took years of discoveries laboriously uncovered by independent investigators who had nothing to go on but their own curiosity and the incongruous information that lay in their paths, challenging them to optimally organize and rigorously refine information that could be formulated into knowledge. These polished concepts which were not in conflict with each other became the disciplines of anatomy and physiology.

It was, in the beginning, not clean work; most of the information was gleaned from battlegrounds and posed by those who were not victorious in gladiatorial combat. The pioneers in these endeavors were often healers or shamans, who inherited from their cultures the received wisdom of mythology (the culturally accepted explanations of natural phenomena; frequently tied closely to religious belief, and supported by what now looks like superstition, but at the time, considered to be divine knowledge, transgression of which often carried dire consequences[1].) There were few trained physicians, since the information necessary for a training program was still being collected and formulated.

In the present, it seems that millennia of investigations, multi-volume books of information, clean laboratories run by investigators in starched lab coats, belie the dirty battlefields where blood has turned the soil to mud, and the accidents that occurred on foggy cobblestoned streets with sewer water running in the roadway gutters. However, a busy trauma center still has its share of gore, and

---

[1] See Plato's drama *THE APOLOGY* describing Socrates' trial and execution for his efforts.

the most pristine laboratories still are arrested with mental conundrums awaiting solution by the kinds of minds that are not stopped by conventional wisdom or suspended by dualities.

Paradoxes, as they were recognized, were set aside, and the knowledge accepted pragmatically. That is, if it worked it was good enough. Science is like that today. Euclidean geometry does not accurately describe the universe in terms of the axioms of relativity, but it was good enough to get us to the moon and back. The disciplines we focus on now are branches of knowledge focusing on the structure of the body (*anatomy*) and the function of the organs that make up the structure (*physiology*). The fields of A &P are classes of information, which, as a group, are comparable to what we have termed Ideal or Abstract, though the individual bits of information are Real or Concrete.

In ancient times, most people distinguished what was out there and what was in here. The enigma began with learning what the organs were for.

HISTORY of ORGAN FUNCTIONS

A surgical papyrus, written c. 1500 BC, was a compilation of 48 cases, including 27 head injuries. In 1862, it was purchased in Luxor by Edwin Smith (1822 – 1906) , after whom it was named. The Edwin Smith Surgical Papyrus is in the Historical Collection of The New York Academy of Medicine. This document is a reflection on the attention Egyptian investigators paid to unknown information.

Hippocrates (c. 460 BC – c. 370 BC) is accepted as the first person to consider disease to be the result of natural phenomena, rather than the will of the gods or supernatural effects. He is, therefore considered to be the *Father of Modern Medicine*; his knowledge of anatomy and physiology was limited by the ancient Greek taboo on dissection of the human body, and his forte was diagnosis and prognosis (understanding the natural course of disease).

Galen (130-200) was the authority for almost two millennia, based on his dissections of monkeys and pigs. As a clinician, he was physician to gladiators and acquired extensive experience on losses of mental abilities secondary to brain injuries. His authority was upended by Vesalius (1514-1564) who published anatomic drawings *De humani corporis fabrica* in 1543. Galen was overturned by Harvey in 1628 when he published the physiology of the circulatory system, *De motu cordis*. Ah well, the history of science is one correction after another.

BRAIN and its FUNCTIONS

And it still goes on. It The Edwin Smith Surgical Papyrus (above) became the first manuscript associating head injuries and brain damage, to loss of mental functions. Plato (c 429-348 BC) claimed that the brain was responsible for mental activities, Aristotle (384-322 BC) felt the brain cooled the blood and the heart was the site of thought and sensation. The connection between brain injuries secondary to strokes informed Broca's (1824-1880) and Wernicke's (1848-1905) studies of speech functions in respective areas of the brain. Wilder Penfield (1891-1976), using stimulation techniques during surgery, defined the motor and sensory areas of the cortex. Franz Josef Gall (1758-1828) defined the occipital cortex as the primary site of visual sensation, an extraordinary bit of neuroanatomy and neurophysiology, which is almost forgotten in his laughable persistence as the founder of phrenology. Remarkably, despite its apparent certainty, science evolves from relative truths. This statement is exemplified by all the convictions above which evolve into a discipline of such reliability. Step by step, we get closer to something in which we can have greater and greater faith.

As time went on, we became more and more convinced that parts of the brain could be defined by their physiologic function. It became urban myth that humans only use 10% of their brains. In fact, the brain is extraordinarily active all the time. It would be unusual if it weren't. The heart, lungs, kidneys and liver work 24/7/365, with the exception of diurnal variation. The 10% myth is based on the fact that parts of the cortex can be damaged without apparent loss of mental function. Check out the story of Phineas P. Gage (1823-1860). His sad story belies the falsehood hidden in the qualifier *apparent.*

Gage worked in Vermont as a construction supervisor on the Rutland and Burlington Railroad. He was an affable, friendly and reliable young man who was blasting rock near Cavendish, VT. on Sept. 13, 1848, when, following a distraction, tamped blasting powder without the requisite layer of cushioning sand. The powder exploded and drove the 13 pound, 3'7" tamping rod through his cheek and brain, landing almost 25 feet away, covered with blood and brain tissue. He not only survived, but healed, with the exception of a personality change resulting in his becoming fitful, irreverent and profane[2].

He lived another 12 years and his skull in the Warren Anatomical Museum at Harvard Medical School in Boston was compared with CTs of men of comparable size and age. The finding was that his brain lost about 4% of its cortical tissue in the frontal lobe. The truth is that the allegedly quiescent cortex is known to cognoscenti as association cortex and remains busy behind the scenes, maintaining memory, comparing notes (hence the name *association*), and carrying out feedback phenomena.

Attempts to understand mental functioning have proceeded both from the top-down; as well as from the bottom-up. The top-down approach is similar to deductive reasoning, in that it begins with a general premise and reaches a specific conclusion. In science, deductive reasoning starts with a hierarchy of

---

[2] Thomas H. Maugh II . Los Angeles Times, May 16, 2012
Damasio, Antonio *Descartes' Error,* 1994, Grosset/Putnam.

simple truths, which are built into a general statement. Developing a deductive approach requires getting as much information as possible on the topic. A literature search, reviewing conclusions that have been accepted by previous researchers helps give a better idea of the general premise into which specific ideas can be expected to fit.

Inductive reasoning is the opposite and begins with specific statements, and ends with the general proposition. Inductive research usually starts when an investigator notes a phenomenon that does not fit into the big picture; collating this eccentric information with other specific observations that allow construction of a big picture that may be modified by the incongruous detail.

The first (deductive, top-down) may be thought of as deconstructing the big picture into its component parts; the second (inductive, bottom-up) as synthesizing subsystems into a comprehensive view. We do not have an understanding of all systems, though neurophysiology and psychological research is contributing mightily to formulating this field. A comprehensive picture is being built in the creative minds who work in this field; being held back by an inadequate overview from work now being done in complexity theory, which probably holds greater claim to understanding the ineffable aspects of dualism - a philosophic name for the mind-body problem.

Those are all modern words for an ancient disagreement. One that goes back at least as far as the controversy engendered between Plato and Aristotle on their concepts of the Soul. Needless to say, much ink has been spilt over the years in disciplines who have tried to clarify an indescribable chasm between abstract and concrete concepts. Many vocabularies have been devised, many synonyms created, entire dictionaries formulated in attempts to illuminate the dark recesses of our minds.

Rene Descartes (1596-1650) is said to have introduced the concept of a separate mind and body. Actually, Dualism is found embodied in many works preceding Descartes. He believed that all mental functions were located in the pineal gland.

Zarathustra (c. 6000 B.C.), Persian philosopher, believed in the opposing forces of Ahura Mazda (Illuminating Wisdom) and Angra Mainyu (Destructive Spirit), both uncreated, acted as moral opposites. Egyptian Religious beliefs contrasted the gods Set (disorder, death) and Osiris (order, life). Ancient Mesopotamian poetry sets Gilgamesh against Enkidu. It's religion contrasts Marduk, the creator against the abyss. The rulers (kings of mesopotamia, pharos of egypt) considered themselves to be divine as opposed to the population.

Plato, (c 429-348 BC), the Greek philosopher, despite his belief that the brain was responsible for mental activities, had a dualistic view of reality representing what we consider to be concrete as REAL, and set it aside from its absolute representation as IDEAL, which could not be empirically tested and was perceived as *forms*.

Meanwhile in the Far East, Hinduism, Zen Buddhism, and Sufism all tend to focus on dualism in a more fundamental way, perceiving the self as different from the rest of the world. In these religious traditions, enlightenment is seen as transcendence of the dualism. In Chinese philosophy, dualism seems much less contentious than in Western countries. Chinese attitude toward dualism can be described less as polar opposites, and more as a reciprocal interaction, allowing stabilization, much as the Autonomic Nervous System works in balancing the sympathetic and parasympathetic physiologies.

Indian philosophy sees two realities, one the range of the gods, the other the shadow of the first. Sounds like Plato had communications with this group. Indian Buddhists, like Dharmakirti held that dualism was between states of consciousness (non-physical) and basic building blocks, not considered any more physical than particles described in quantum physics as points of energy. Pity that he wrote in the 4th century. He might have been on to something. At any rate, it would have been interesting to see how he would have formulated his thinking if the information from quantum physics had been available.

Before understanding is accomplished, many more attempts from both the synthetic and from the deconstruction camp will be tendered and just as many undertakings will attempt to find common ground between the two bivouacs. not only has physiology, psychology, and other aspects of brain studies laid claim to bottom-up research, but they have closed shop by requiring post-doctoral studies in their areas. Likewise, Artificial Intelligence, Linguistics, and philosophy continue to work from the top-down area. There is no old-fashioned avocation comparable to what used to be called natural philosophy from which good minds could attempt to anneal the areas of difference. Natural philosophy was the forerunner of natural science; and concerned itself with questions which had not yet been established by the scientific method. Information acquired after the Scientific Revolution is usually considered Natural Science.

This work will try to affirm the primary business of philosophy, which in the words of Heidegger[3], "is to preserve the force of the most elemental words...". Instead of further confounding the thesaurus, I hope we simplify the forest trails.

Simple Definitions

Let us begin, first of all, by distinguishing the mind from the brain. **The brain** is the main organ of the central nervous system (the system of organs responsible for communication). The brain transmits impulses throughout the body, by several means:

1. by transmission along the nervous system
2. by feedback via the autonomic nervous system
3. by releasing information via memory from association cortices
4. by hormone release through the blood stream; and across synapses

---

[3] *Being and Time*. 1962,Harper & Row. MacQuarrie, Robinson translation of SEIN und ZEIT.

The brain is a concrete structure, and the transmissions of impulses listed above are by means of concrete mechanisms; electron transmission along the neurons, electrolyte transfer and molecular flooding of the synapses in 1, 2 and 3; and hormone molecules in 4.

This communication system is involuntary, which means we do not need to make a conscious choice to make it work. There are many places conscious choice is necessary (e.g., motor activity, thinking), and many places where conscious choice is outright inconvenient (e.g., the maintenance of pulse, respiration, and monitoring of blood pressure). Indeed, those functions which the wisdom that is evolution has relegated to unconscious function has become part of autonomic functioning and emerged as part of a complex scheme involving feedback, so those systems which need to be monitored are automatically controlled; speeded up and slowed down when necessary to maintain balance. This is known as homeostasis, and will be discussed more fully in its proper place.

When talking of the brain, the words *conscious, voluntary,* and *aware* are synonyms. As is usual, in English, synonyms[4] do not necessarily mean the same thing, but can be used interchangeably for the same sense. The synonyms for AWARE are: *COGNIZANT, CONSCIOUS, SENSIBLE, ALIVE,* and *AWAKE*

- AWARE -- may indicate either general information, wide knowledge, interpretative power, or vigilant perception
- COGNIZANT -- may imply the gradual impingement of knowledge or perception on one's consciousness or may connote special efforts to know
- CONSCIOUS -- may indicate impingement on one's mind so that one recognizes the fact or existence of something
- SENSIBLE -- may apply to situations in which a thing is intuitively sensed and also to those in which it is rationally perceived, known, and admitted
- ALIVE -- may suggest vivid awareness, certain keen perception
- AWAKE -- may suggest alert perception

The synonyms for *voluntary* are in a separate category, seeming to dwell in a different dimension of thought. The *Aware* or *Conscious* group of synonyms can be voluntary or involuntary, but the *Voluntary* group of synonyms all include things we choose to act on as a result of awareness or consciousness. We will see, as we explore the categories of definitions, that the activities that encompass voluntary things are not only in a separate category, but also require precedents from a different dimension (involving the *will*, a subcategory of mental activity). The synonyms for VOLUNTARY are: *INTENTIONAL, DELIBERATE, WILLFUL,* and *WILLING*

- VOLUNTARY -- implies freedom from any compulsion that could constrain one's choice; often it suggests merely spontaneity, or, in contrast with involuntary, stresses the control of the will
- INTENTIONAL -- contrasts with accidental and inadvertent in specifying an intention and purpose
- DELIBERATE -- carries the idea of full knowledge or full consciousness of the nature of an intended action
- WILLFUL -- adds to DELIBERATE the idea of a refusal to be advised or directed in any way and an obstinate determination to act despite all wiser opposing forces or considerations
- WILLING -- implies such qualities as agreeableness or openmindedness that make one ready or eager to accede to others' wishes or effect an end pleasing to them

---

4 Merriam-Webster's unabridged dictionary, version 3:

We are aware of some of the things that the brain does and we are not aware of other things that the brain does. The most obvious things in the first category are the things registered into the parts of the brain called the primary cortices.The best known primary centers are the visual cortex, somatosensory cortex, auditory cortex and motor cortex.

The visual cortex lies in the occipital lobe (V1 & striate cortex)(Brodman's Areas 17-primary, v1,18secondary visual cortex, v2,19 - associative visual cortex, v3,4,5)

The somatosensory cortex lies posterior to the Central Gyrus in the parietal lobe (Brodmann's area 1,2,3)

The motor cortex lies anterior to the Central Gyrus, in the frontal lobe (Brodman's area 4, secondary 6)

The auditory cortex lies in the temporal lobe (Brodman's areas 41&42)

The mind, on the other hand is an abstract function of the brain, as opposed to what we have called above the concrete functions of the brain(i.e., transmission of information). This distinction which we have just made about concrete and abstract functions of an organ brings to mind one aspect dwelt on by philosophers of consciousness, namely, epiphenomena[5]. In the case of the brain, the phenomena referred to is the physical physiology of the brain (transmission of sensory impulses through the primary and secondary sensory organs to the brain, transmission of decisions made by the brain through the motor nerves to the musculoskeletal system). In contrast, the epiphenomena is what happens between the stimulation of the sensory system and the response of the motor system.The mind works by methods not yet understood, and may be described as the awareness the organism has of the brain's activities. There are many postulates in the philosophy of science to help explain how the mind occurs.

---

[5] a secondary phenomenon accompanying another phenomenon and thought of as caused by it. Merriam-Webster's Unabridged Dictionary, version 3.

dualism -- there are two elements of existence: Physical and Spiritual. mind and matter are two ontologically[6] separate categories. Dualism can refer to moral dualism, (e.g. the conflict between good and evil), mind-body or mind-matter dualism claims that neither the mind nor matter can be reduced to each other in any way.(e.g. Cartesian Dualism) or physical dualism (e.g. the Chinese Yin and Yang).

monism -- a philosophical position which argues that the variety of existing things can be explained in terms of a single reality or substance.

idealism -- The group of philosophies which assert that reality, or reality as we can know it, is fundamentally mental, mentally constructed, or otherwise immaterial.

materialism -- the only thing that exists is matter or energy; that all things are composed of *material* and all phenomena (including consciousness) are the result of material interactions. In other words, matter is the only substance, and reality is identical with the actually occurring states of energy and matter.

Pluralism is a term used in philosophy, meaning "doctrine of multiplicity", often used in opposition to *monism* ("doctrine of unity") and *dualism* ("doctrine of duality"). The term has different meanings in metaphysics and epistemology.

In metaphysics, pluralism is a doctrine that many basic substances make up reality, while monism holds existence to be a single substance, often either matter (materialism) or mind (idealism), and dualism believes two substances, such as matter and mind, to be necessary.

In epistemology, pluralism is the position that there is not one consistent set of truths about the world, but rather many. Often this is associated with pragmatism and conceptual and cultural relativism.

It is thought by many neuroscientists that those methods by which the mind works will probably be similar to concrete mechanisms similar to those employed in brain physiology. Some of the work done in PET scanning already suggests this.

----

[6] Derived from the science or study of being; a theory concerning the kinds of abstract entities that are to be admitted to a language system. Merriam-Webster's Unabridged Dictionary, version 3.

Let us continue, secondly, by clearing some brush[7] from the path:

Mind, is defined[8] in thirteen different ways, some in direct opposition to others. The following selection forms a representative choice:

- *memory* -- as in re-mind
- *intellect* -- especially in opposition to will and emotion
- *response -- an organized group of events in neural tissue occurring mediately in response to antecedent intra-psychic or extra-psychic events which it perceives, classifies, transforms, and coordinates prior to initiating action whose consequences are foreseeable to the extent of available information*

The variety of definitions of the mind have to do with selecting from the myriad of things which go on in one's head; the collection of thoughts that go on in one's mind forms one of the busiest places an individual can visit.

The aspects of the mind are conscious and unconscious. It is not clear yet what the difference is between the two. Superficially, conscious mental activities are those of which we are aware, implying a degree of control over conscious activities. It would seem that the opposite is true - i.e., that unconscious activities are those of which we are unaware - but it's not that simple. It seems intuitive to say we are aware of conscious activities and unaware of unconscious activities. Already, we are running into paradoxes, as dreams come from the unconscious, but enter our awareness. We are often aware of unconscious activity, such as dreaming, and occasionally aware of other unconscious activity, such as the activity or conversations that happen around us while anesthetized for surgery, and even rarely hear of people in coma reporting things they have been told while insensible. Furthermore we have all heard of people who return from clinical death, who recall activity in the room while they are being resuscitated.

---

[7] with the caution, ala Kant, that what is left after the pruning is the creation of my mind. No apologies. It worked well for me and may be of use to others. That's the reason for writing this and any other creations.

[8] Merriam - Webster's Unabridged Dictionary, version 3

In addition there is the term *unconscious[9] mind,* as understood in psychoanalytic theory, a term that was coined by Friedrich Schelling (1775–1854), a German romantic philosopher, introduced into English by Samuel Taylor Coleridge (1772 –1834), English poet and essayist, and developed and popularized as a concept by Sigmund Freud (1856-1939), Austrian neurologist and founder of psychoanalysis

Empirical evidence suggests that unconscious phenomena include:

1. repressed feelings -- expressed as dreams in a symbolical form
2. automatic skills, reactions -- expressed as slips of the tongue
3. subliminal perceptions -- expressed as jokes
4. thoughts
5. habits, complexes, hidden phobias and desires

In psychoanalytic theory, the unconscious mind can be seen as the source of dreams, the origin of automatic thoughts, with no apparent source, and as a well of forgotten memories (that may still be accessible to consciousness at some later time). In all, it can be interpreted as a locus of implicit knowledge (the things that we have learned so well that we do them without thinking). The influence of the unconscious mind is so extensive that it could easily be interpreted as a modern substitute for what has been called *a priori* knowledge in the past. Indeed, it may well be Socrates was referring to the slave in the *Meno,* which Plato interpreted as knowledge brought from a prior lifetime.

It has been argued that <u>consciousness</u> is influenced by other parts of the <u>mind</u>. These include <u>unconsciousness</u> as a <u>personal habit</u>, <u>being unaware</u>, and <u>intuition</u>. Terms related to semi-consciousness include: <u>awakening</u>, <u>implicit memory</u>, <u>subliminal messages</u>, <u>trances</u>, <u>hypnagogia</u>, and <u>hypnosis</u>. While <u>sleep</u>, <u>sleep walking</u>, dreaming, <u>delirium</u> and <u>comas</u> may signal the presence of unconscious processes, these processes are not the unconscious mind itself, but rather symptoms. Being aware sometimes happens when we are not conscious, i.e., when we are sleeping and dreaming, we are in a state not usually classed as conscious; but with some effort we can retrieve the information. Along the same lines, the literature is full of examples of scientific discoveries that have been resolved in dream states.

Some of the distinctive examples are summarized by Dr. Deirdre Barrett[10] of Harvard Medical School:

John Steinbeck - "It is a common experience that a problem difficult at night is resolved in the morning after the committee of sleep has worked on it."

---

[9] not to be confused with loss of consciousness, although both are states in which awareness is lacking, and may share more basic characteristics, when full understanding is accomplished.

[10] Barrett, D.L. The 'Committee of Sleep': A Study of Dream Incubation for Problem Solving. Dreaming: J. of the Association for the Study of Dreams, 1993, 3, p. 115 123.

August Kekulé - the chemist reported that his Nobel-prize winning realization of the structure of the benzene molecule as hexagonal rather than straight came after dreaming of a snake grasping its tail in its mouth.

Dmitri Mendeleev - described dreaming the periodic table of the elements in its completed form (Kedrov, 1957, pp. 91-113).

Otto Loewi - The Nobel-prize winning experiment demonstrating the chemical transmission of nerve impulses to a frog's heart was conceived in a dream (Dement, 1974, p. 98).

Elias Howe - sewing machine needle—with the hole at the pointed end

J. B. Parkinson - computer-controlled anti-aircraft gun.

William Blake - described being told by his dead brother in a dream about a new way to engrave his illustrated songs.

Samuel Taylor Coleridge - states in the preface to *Kubla Khan* that the poem appeared complete in an opium-induced dream.

Robert Lewis Stevenson - dreamed the two key scenes of his novel, *Dr. Jekyll and Mr. Hyde*.

Giuseppe Tartini - composer heard *Devil's Trill* in dreams.

Igor Stravinsky - composer heard *Rite of Spring* in dreams.

Jack Nicklaus - credited a crucial improvement in his golf game to dreaming of a new way to grasp his club.

Clearly much of what is stored in memory in the conscious life is available to the dream time, and vice versa, much dream time learning is available, albeit not readily, but with effort, to the awake time. Also, when we forget something, we are not aware of that thought, but are conscious[11].

One of the things that confounds the above discussion is that the definition of *consciousness* still lies in a nebulous area, as acknowledged in footnote 7. In the old days, and in the current Webster definition, consciousness means awareness or perception.

Awareness is the state of being aware (knowing what's going on).

Perception is a state created by our brain due to stimulation of our sensory nerves. In neurophysiology, when the light from a landscape hits the retina at the back of our eyes, we are **stimulated**; when that stimulation travels back the optic nerve, crosses the optic chiasm, registers at the lateral geniculate, and travels down the optic radiation, registers on the occipital cortex at V1, then V2, V4, and

---

[11] Freud has suggested that these are often items of the greatest importance to our mental hygiene. Brazleton, on the other hand has claimed that (my paraphrase) selective ig-norance is a sign of intelligence.

V5 (MT) is distributed throughout the cortex and is "seen" by us that is *perception*. Visual experience can occur without sensory stimulation, as in dreams, when the digital image, which has been stored in memory, is re-digitized and re-perceived.

Indeed, it is a well know phenomenon in psychology, as well as in Art, that a negative space in the visual field will be filled in by the mind, and a well known fact in jurisprudence that eye witness testimony is not the most reliable.

In our furtive attempt to simplify definitions, we have inadvertently demonstrated the complexity of the synonym bush. Let us take a step back and overview the mind. It is certainly difficult to not try to think like Plato and appreciate the brain (Plato's **Real**) and it's function, the Mind (Plato's **Ideal**). To see this as dualism is, in a way, banishing a complex relationship to a duplex existence with their dependency forever cursed in dualistic antagonism, as adversaries, who can, by nature, avoid the common ground that forms the basis of their existence.

So, we will concentrate on developing the Ideal, and hope, by exploring this ground, to offer a way of looking at the whole from one side, anticipating that the negative space gets filled in.

The traditional Mind, in Psychology, usually is divisible into cognition, emotion, and will. These divisions of the Mind are amongst the earliest definitions that have been proposed by introspective investigators to represent different functions that have been perceived in the early considerations of mental functioning. Later investigators, often with a religious base, have proposed definitions which were accepted as though they came from on high. The vocabulary of mental functioning expanded, often without empirical evidence.

Our attempt will be to explore those definitions to the greatest and most authentic meaning possible in the service of providing a base upon which future evidence can be constructed, or which future evidence can justify or negate, as the case may be. The definitions which will form axioms for our foundation concepts of the mind are:

Divisions of the Mind
>Intellect
>Emotion
>Volition

Steps in the process
>Sequential components comprising processes within Divisions, as
>>described in the chapters on individual Divisions.

Boxed definitions, where indicated by footnote are derived from dictionaries or standard texts. Otherwise they are explanatory based on the context. There are many ways of dividing the multiple functions the mind can be construed as having, e.g.:

The three major functions of the Soul (psyche), according to Plato[12], are Rational (Intellect), Spirited (Volition), and Appetitive (Emotion). These are his assessment of how the Mind works, and form the analogy of how a society functions. The Rational part has the function of governing, is analagous to the head of the body, and is symbolized in society by the Rulers or Philosopher King. The Spirited part has the function of protection is analagous to the chest of the body, and is symbolized in sociey by the Warriors or Guardians. The emotional part has the function of production, is analagous to the abdomen of the body, and is symbolized in society by the Workers.

Intellect, Emotion and Volition are always working, and if functioning smoothly, a good balance (homeostasis[13]) is maintained. The effortless interaction amongst these activities seems to occur when we "let it be.[14]"

The brain, as the physical part of living,is a concrete entity. The para-physical, or abstract, Mind[15], is necessarily dependent upon the brain; mind does not exist without the brain. It is true that brain is not sufficient for mind's well being. There must be input, such as sense-perception, experience, challenge; and output: action, ideas, creativity, for the mind to function. The capability of the mind to process these is character; the aura, atmosphere, or gestalt is personality: the interaction, growth. The influence of these things on the environment is spirit – and that certainly lives on.

The need for the mind to not only function, but cause action, is the final arbiter of its health. Being without Doing is a kind of paralysis – a living death. The interplay of being and doing maintains health and promotes growth if there is consistency of Will, Emotion, Intellect, and Action. Otherwise the malfunctioning of the system may worsen in a spiral fashion.   When the parts of the mind do not interact harmoniously, there are usually two ineffectual parts causing competition, rather than cooperation; and a third well functioning part which takes on the role of referee; for example:

If the Will and Emotions are at war, the Intellect may arbitrate, leading to dispassion, coldness and automatism. This is usually the result of mistrusting the emotions, an overactivity of a mal-developed will, a perverted conscience. The solution is to allow the emotions to interact with the authentic will. It is in the nature of the mind that a homeostasis will be reached in which the will and emotions will interact in a productive way. The fear that society has of one running rampant on pure emotion leads to an attempt to over-control the

---

[12] *The Republic*

[13] A term coined by Walter Cannon in his book *The Wisdom of the Body*. It refers to an interlocking series of feedback mechanisms, which assure that balance is kept in biologic systems

[14] Apologies to Paul McCartney

[15] This aspect has been introduced in the discussion of Abstract categories of Existents in *Meaning,* by Richard Stevko, Graven Image Publishing.

emotions by Will or Intellect. Whichever one is not engaged in battle with Passion becomes the arbiter. A homeostasis has been reached, but not a productive one. So the lack of faith in natural homeostasis leads to balance at another level –unnatural and unproductive.

It is the same if the Intellect and Emotions battle. This is caused by thinking we can "figure out" a way of life which is better than what our innards[16] tell us. Here the Will becomes so engaged in refereeing, that it is immobilized from making any practical demands on action.

A battle between Intellect and Will, between what we know we want and what we want to know, will be arbitrated by passion. This leads to pleasure seeking without commitment or purpose.

A Broader Look at Homeostasis

This preliminary formulation of aberrant homeostasis seems to be the result of attempted solutions turning back on the problem. It is a more definitive statement of the problem which occurs too commonly in which the Will shapes or is shaped by Intellect and Emotion. These tentative statements are demanding further exploration, which will be done when we understand the factors a little better.

How do these anti-productive patterns develop, if there is such a homeostatic urge for natural balance and interaction in the mind? They seem to develop from a lack of faith in homeostasis. A belief that "letting it be" will not work out in our best interests. One way of seeing how that happens is to consider the newborn child. It has been expelled from its environment, cries lustily, arms and legs thrashing and thrusting. We say it doesn't like the world we've kept. So true. The baby's will is to not be here – the choice is to object. The intellect recognizes a new environment, and actions consistent with the emotion of rage are expressed.

Now, many would say, "There's no Will in a newborn. There's no Intellect. They don't know or choose anything this early. What you see is pure unadulterated emotion." They are certainly right to say the infant does not rationalize and choose the way adults do. The infantile intellect is simply to know things are different. Not visualize or conceptualize in an adult manner, but know now that there is a world outside the womb. The child also does not choose to cry. The Will has neither precursors of intent nor goals of expectation. The most that can be said is that the baby cries rather than not crying. There are, after all, healthy infants who cry little, some not at all, and many only briefly. To object strenuously or not is Will, or at least the closest to what can be called Will. If the

---

[16]This colloquial substitution is meant to avoid the accumulated meanings heaped upon the word *Intuition* by overuse of many disciplines, but is meant to refer to the accumulated internal response sensed by the organism.

child's development is observed dispassionately, that quality can be seen to develop, before age two, into what everyone would call "willfulness", the NO stage.

The three functions of Mind: Intellect, Emotion, and Will, are congruent. The baby knows the environment has changed, is uncomfortable, and wishes it to be not so. If the child is left alone, in effect told, "your innards don't know what they're talking about", the child will deteriorate. Even if hunger and exposure are prevented the end is tragic. At the other extreme, if the uterine conditions are reproduced; warm, wet, dark, silence, and then the homeostatic conditions are returned to a fetal level, no growth occurs[17].

But neither is done for very long. The child cannot be taken back to where it came from, despite the avowed desires in that direction of older siblings. It must live in an extrauterine environment. Nor is it left alone. Maternal desires to hold, cuddle, and cozy the baby partially reproduces the prenatal environment resulting in new learning for the child, comfort of the rage, and gradual decision that this new world is OK, too.

What has been observed is not hunger – most babies eat little or nothing for a day or two. It is not discomfort – many efforts are made to relieve cold, pressure, pain of any sort. It is simply a new experience, a change of environment, and the child first demonstrates an innate activity of the three mental functions, and then reaches a homeostatic balance.

Most of the adult responses to this situation are, and remain for awhile, as primitive, innate and animalistic – that is, as well balanced – as the child's initial outburst and acceptance. As has been shown, to be otherwise would have drastic consequences. Where, then does the lack of faith in homeostasis begin? Obviously, it begins at different times for different people. Here are some common patterns. They are oversimplified. Human interaction is so complex and individual that all aspects cannot be included – even in describing a single specific case. But the generalizations contain a wide variety of specifics, and can be applied broadly, but not universally.

Consider a child of about six months of age who has previously slept well, and begins awakening five and six times a night. It is known that separation fear begins at this age, with the child's emerging individuality accompanied by fear of maternal gone-ness. An appropriate behavior, in this society, would be to answer that fear by making an appearance, letting the child know that it is individual, but not alone. The homeostasis of intellectual realization (knowing mother is there) and of willful acceptance (going back to sleep) is reestablished.

If the fear is given priority (affirmation of emotion, lack of faith in Intellect or Will) and the child excessively consoled, such as always being taken to bed with the parents when it cries, then the emotions predominate, and the child uses no

---

[17] Spitz, R.A. (1945). Hospitalism—An Inquiry Into the Genesis of Psychiatric Conditions in Early Childhood. Psychoanalytic Study of the Child, 1, 53-74

knowledge or Will in a stress situation – it always cries. This example can be multiplied endlessly over the ensuing years, with eventual counterbalance of emotion with Will or Intellect, and eventual over-control or arbitration by the other.

Other examples can be cited of affirmation of either Will or Intellect and resultant engagement of the other two in either opponent or referee positions. Case histories abound –talk to any parent. The explicit reference to mental homeostasis may be absent but can easily be seen.

How can homeostasis be restored? Whichever way it is restored requires an identification of the types of imbalance. In any situation, especially one in which you feel guilt, anxiety or just that something's not right – think of why you acted the way you did. This "reason for action" is usually the doorway into the Will. If the answer to "Why did I act the way I did?" always seems to come up "because it was right", then you have an overactive will. If it really was the right thing to do or way to be, then there would be no guilt, anxiety, or feeling that things aren't right. If this is your answer, then the next step is to decide what the Will is warring against.

There are many possible combinations. An effective exploration must at this point face the problem which has twice already thrust itself in, what are the interactions of Intellect, Emotion, and Will?

THE COMPONENTS OF MENTAL FUNCTIONING

The possibilities of interaction amongst Intellect, Emotion, and Volition alluded to at the end of the preceding section, can be tabulated:

Dominant	Opponent	Referee	Manifestation
Intellect	Emotion	Volition	Willfulness
Emotion	Intellect	Volition	Willfulness
Volition	Emotion	Intellect	Automatism
Emotion	Volition	Intellect	Automatism
Intellect	Volition	Emotion	Hedonism
Volition	Intellect	Emotion	Hedonism

It should be obvious that the manifestations are based on the assumption of a referee stronger than the other parameters. It is just as possible that the referee may be immobilized under the strength of either the dominant component or the opponent. The issue is apparently more complicated than the outline.

What has, in effect, been set is a stage for six mini-dramas, each including a Protagonist (referred to above as 'dominant'), an Antagonist ('Opponent"), and a Denouement ('referee'). Although the conflict is set for each play, the denouement may be reached by the referee interacting with each character in an a) affirmative, b) acknowledging, c) ignoring, d) disavowing, or e) negating way. This results in twenty-three possible outcomes for each drama.

An example might be:

Emotion: "I want a candy bar."
Intellect: "It'll ruin your appetite."
Volition: makes one of the following responses:

1. affirms emotion + negates intellect= "I will do it and nothing can stop me."

When Volition affirms emotion, essentially saying "you should have a candy bar", and negates intellect, essentially saying, "It won't ruin your appetite"; then there is agreement to the appetite and no objection to the candy bar and it will be acquired and nothing can stop the snack.

2. ignores emotion + negates intellect = "Nothing can stop me."
3. affirms emotion + disavows intellect = "I will do it; no reason not to."
4. affirms emotion + ignores intellect = "I will do it."
5. affirms emotion + acknowledges intellect = "I will do it, but shouldn't"
6. acknowledges emotion + negates intellect = "I want to, nothing can stop me."
7. acknowledges emotion + disavows intellect = "I want to; no reason why not."
8. acknowledges emotion + ignores intellect = "I want to."
9. ignores emotion + disavows intellect = "No reason not to."
10. ignores emotion + acknowledges intellect = "I shouldn't."
11. disavows emotion + ignores intellect = "I don't want to."
12. disavows emotion + acknowledges intellect = "I shouldn't, and don't want to."
13. negates emotion + acknowledges intellect = "I shouldn't; nothing can make me."
14. acknowledges emotion + affirms intellect = "I want to, but won't."
15. ignores emotion + affirms intellect = "I won't do it."
16. disavows emotion + affirms intellect = "I won't do it and don't want to."
17. negates emotion + ignores intellect = "Nothing can make me do it."
18. negates emotion + affirms intellect = "I won't do it and nothing can make me."
19. negates emotion + negates intellect = "Nothing can make me or stop me."
20. affirms emotion + affirms intellect = "I will do it and I won't do it."
21. acknowledges emotion + acknowledges intellect = "I want to, but shouldn't."
22. disavows emotion + disavows intellect = "I don't want to, but no reason not to."
23. ignores emotion + ignores intellect = "…"

The first nine responses show intent to get the candy bar, in decreasing order of strength; the next nine indicate intent to not get it, in increasing order of strength; the remaining five are conflict situations of decreasing intensity, and resultant action is in question.

Each situation is actually a conflict situation. Indeed, the conflict remains the same; wanting the candy bar, knowing it will ruin the appetite. It is the resolution that produces secondary conflicts, or rather, accentuates the primary conflict. There seems to be no way out. It is impossible to affirm both Emotion and Intellect. The next best solution, the closest to mutual affirmation, is to acknowledge both (#21), but if the final action is to get the candy, Emotion is affirmed and Intellect negated (#1); if it is not gotten, Intellect is affirmed and Emotion negated (#18). It would seem satisfactory to get the candy, and save it until after dinner. That way both sides of the conflict are affirmed. Compromise, however, is good in politics, acceptable in inconsequential matters of a single candy bar, and counterproductive in substantial decisions of the mind. How, for example, can one compromise in ultimate decisions such as killing in war. It is true that such compromises are made all the time, but the mental processes are not congruent, the action is not authentic, and the resultant secondary conflict, or accentuation of the primary; conflict is difficult to live with. At the very least, such compromises reinforce one of the participating mental processes to such a degree that one's freedom is henceforth restricted to the degree that the reinforcement is accepted.

It is entirely possible, and even likely, that any one of the twenty three above solutions represents a habit pattern and acquiescence of Will to whichever of the other two mental functions has been previously reinforced, tempered by the degree of strength of the other. This is the idea of arranging solution #1-18 in that order to represent the varying degrees of strength between Emotion and Intellect. If they are equally strong, solution 19-23 represent the immobility of will which results.

Though these are the most common methods of resolving mental conflict, the issue here is not effective compromise, but authentic behavior. Action, which is the result of mental congruity, is authentic behavior when it is a productive interaction of Intellect, Emotion, and Will. To this end, it does not seem adequate to deal with each mental partition as a discreet whole, but each step in the processes needs to be partitioned just as mind itself was partitioned[18], examined at each step, and ultimately reintegrated.

---

[18] See chapter on Categories of Existents in MEANING by Richard Stevko, Graven Image Publishers.

INTELLECT

INTELLECT

The following definitions will serve to frame the boundaries of what is meant by Intellect, and its contained implications:

Intellect – that division of the mind which knows, reasons, judges, understands, or comprehends. Differentiated from, but related to Emotion and Will. The ways in which these three – Intellect, Emotion and Will are integrated and influence each other will be covered in the section on *Integration of Mental Function*, and expanded in the section on *Interaction of Mental Functioning*.

Intelligence – the power of intellect.

Cognition – the act in which intellect engages, (thinking)

Thought – the subject of cognition.

Idea – the object of cognition.

Thinking (a less formal and more widely used term than Cognition or Cogitation) as an act can be represented by various stages in its progress. First, there are several means of Thinking:

Consider—to fix ones mind on something.

Reason –consecutive, logical thought.

Deliberation – slow and careful reasoning.

Secondly, there are modes of Thinking:

Ponder – a general term for careful thought. (Muse)

Study – ordered (usually linear) pondering. (Meditate)

Weigh – pondering more than one thing (relatedness). (Contemplate)

Revolve – pondering all sides. (Ruminate)

The terms in parentheses following the definition indicate a lesser rational degree of the original term, but a more holistic involvement.

Thirdly, Thinking can be related to time:

Remember – casting thought back to the past.

Reflect – turning thought on itself in the present

Speculate –casting thought into the future.

All of the above definitions are necessary to a process of thought, but obviously are not sufficient. The process begins with Ignorance[19]; ends with Knowing[20]. More accurately, it begins in a primitive state of knowledge, a state of receptiveness better described as Alertness.

A broad overview of the steps (bold-faced) which occur in the Intellect as it progresses from **Alertness** to **Knowledge** begins when Intellect becomes **Aware** of its own ignorance or **Aware** of Information which it does not possess. The two kinds of Awareness are respectively subjective (aware of its own ignorance) or objective (aware of information it does not possess); but is a state of **Awareness**, nevertheless. Once **Aware**, the Intellect develops a state in which it becomes receptive to incorporating the information. This receptivity is called **Wish**. Developing the Wish to possess the information is an Intellectual (the Emotional counterpart to Wish is Desire) The Intellect turns its attention to the information in an evaluative way, and this direction of attention is called **Regard**.

Now regarding the information, Intellect begins a process of evaluation. Regard (looking at) is followed by more careful evaluation of the information called **Assessment**. The evaluation process becomes more intense progressing from Assessment to **Pondering** to **Study**. These steps have been described above and require repeated feedback to prior information in memory and speculation of possible connections to other bits of information or likelihood of possible connections to other bits of information in memory or possible combinations which have not been experienced but seem plausible.

This evaluation process represents the information at hand and compares it to prior bits of information and connections real or imagined and alerts Intellect to ways in which the information at hand fits in with other information in memory and in imagination. This phase is called **Representation**.

Intellect, having processed the information thus far, and storing in memory, has now possesses the information in a state known as **Knowledge**, and an additional function remains in its own sphere, and that is **Demonstration**. Knowledge and Demonstration are different subjectively within the mind, one being possession and the other expression, but are difficult to tell apart, objectively, by an evaluator outside the realm of the possessing mind. That evaluator can judge that state only when it is demonstrated, either by explanation or manifestation of some sort.

---

[19] Technically, the Mind is never in a state of ignorance. The root of ignorance is *ignore*, an activity of which the mind is incapable. From the time the brain develops, indeed as part of its development, there is information flowing into it. We may later choose not to account for information available to us, but we have never been able to ignore it. This is one of the phenomena Heidegger referred to as the business philosophy has to preserve the force of the most elemental words (p. 12). The ignorance of which we speak in this book is more correctly defined as not knowing.

[20] Knowing, the term as used here refers to a reasonable person's pragmatic knowledge; epistemological justified belief is beyond the scope of this work, although many steps in this process may be found to apply there as well.

Abstractly, the above steps in that process may be grouped as:

1. Precursory Conditions,
2. Consideration,
3. Conception,
4. Evaluation, and
5. Realization (not in the sense of perceiving or conceiving, but making real: real-ization.

Looking more carefully at each of these steps, we find that...

(1) Precursory Conditions – the precursory conditions describe the beginning of thought. The beginning of thought, in its purest form, is Ignorance. This Ignorance is not total, but is a steady state characterized by lack of the specific knowledge impinged upon by Evidence, which intrudes upon it. The intrusion of Evidence requires an Alert state, which causes interaction of the Intellect with Emotion[21], creating Awareness (whether it is an awareness of new information; or an awareness of the prior ignorance). Having become Aware, another emotional response (Desire to incorporate that information into the data bank of the brain) tips Awareness toward Wish. In this state, the sense organ's attention is focused upon the Evidence in a state now recognizable as Regard.

Ignorance —▶ Evidence —▶ Awareness—▶ Wish—▶ Regard
      ↓   ↑   ↑
      Desire —▶ →Attention

Were we able to begin this discussion with an Intellect containing no information, we would have the hypothetical blank slate, a tabula rasa, an empty mind. Biology is never so simple. All beings have innumerable experiences before birth[22]. The mind is never blank. The beginning of the thought process is never one of incogitance, there has always been some kind of thought going on, primitive though it may be. The state of ignorance is always a relative state, depending on the experience of the organism. What we are referring to as Ignorance in this context is a relative term. Ignorance is meant to indicate a state of mind in which information is novel to the consciousness.

---

[21] Although this section concentrates on the process within each component of mental functioning (Intellect, Emotion, Will), the interaction amongst the components of mental functioning is being suspended and will be more closely examined in the section on **Integration of Mental Function.**

[22] The very process of development of the brain is replete with neural signals. Perhaps one of the more elegant of these is the internal signaling of the retinal cells to corresponding visual cortex cells on the extreme opposite side of the brain (eye to occipital cortex) to assure that corresponding points of visual reference are aligned properly. Many more subtle circuits are established as part of the development of architecture of the nervous system. In addition, the brain receives constant input from its sensory systems before birth.

Ignorance is always with us. There are many reasons why one might choose to remain in this state, but they will be explored in the section on abnormal mental functioning. It may seem arbitrary at this point to call that "abnormal", since no one can ever know everything, and it seems foolhardy to call the attempt to do so "normal " – an impossible task at best. What is referred to here is, not knowing everything, but ordering in some cognizable way those things with which we come into contact. That brings up the next precursory condition, Evidence.

Evidence is considered to be those things with which we come into contact, specifically those things which impinge upon us. It may include such things as sense perceptions: vision, hearing, touch, smell, taste; ideas; emotions. The impingement of these evidences may be called Awareness.

Awareness involves more that just the impingement of evidence on ignorance. It is also recognition of ignorance, and also involves emotional and volitional components which will be discussed in those sections and the section on integrated mental functioning.

Wish is the next precursory step. One can remain in a state of awareness only if the impinging evidence is consistent with prior knowledge. Actually it's a very pleasant state to be in, but new evidence, along with emotional and volitional components pushes Awareness to Wish. This might be argued to represent only Desire, an emotion, but is actually the intellectual concomitant or result of Desire. It is a concretization, in thought, of a bridge between Awareness, its predecessor, and Regard, its antecedent.

Regard is simply "looking at". It is the point at which one takes in the impinging evidence, not as recognition of something new, but with intent of doing something with it. That "doing something" is a movement into the next phase of Thinking, Consideration.

(2) Consideration – includes assessment, pondering, and study. It actually has its roots in Precursory Conditions, as those phases are often referred to as a matter "under consideration". This kind of reference is only an acknowledgement of the roots Consideration has in Precursory Conditions.

Assessment, as has been said, is Regard with Intent. It is not sufficient to leave the definition at that, but it's more complicated, involving Emotion as well. It too will be dealt with in the section on integrated mental functioning. Assessment may be further defined as taking in evidence, not only with intent, but in an ordered way, which is often determined by the backward influence of the next state, Pondering.

(3) Conception – is the formation of an idea. It includes Pondering, with all its sub states, and Study. Note that the groups of Consideration and Conception not only overlap, but blend, one into the other. Pondering and Study are in both groups and there is no line of demarcation between them. As Evidence is Pondered, each comparison (referred to as sub-states below). takes it more into knowledge or certifies the information as being already Known material. Here the

raw materials, or precursory conditions of Intellect, are processed and ordered into a realizable whole. We will see similar overlapping in Evaluation and Realization.

Pondering is a highly complex state, including many sub states, including a casting of thought to the past and future as well as the present, and including impingement of both Emotion and Will. This last point is not just an interaction with the other mental states, which actually happens at each stage as will be seen later. It is not only concomitant or comparable states, as has been alluded to already; it is an actual impingement, a drawing in of the processes in the other states of mind. That will become clear only with demonstration.

The central aspect of Pondering is Comparison. The information which has been assessed is now compared with information from the memory (past), and from speculation (future) which may include the emotional components of speculation Imagination and Estimation. The Comparison aspect of pondering may be returned to in the future phases of the intellectual process when the material now being processed reaches the point of Representation and Experiment, being brought back for reconsideration of the results and reassessment. These complicated interactions which cross time barriers will be diagrammed after consideration of the present (Now ness) aspects of Comparison.

What is compared, is correlated in the sense both of being ordered to make sense (organized), and in the sense of data from the past and future being compared to present data. That Correlation is in a way a kind of movement into the future -- it has the sense, or is ordered by, what could/should be. There is a sense of expectation in it. The correlation is then moved back to the level of Assessment, for Discrimination. This process involves taking back what has been correlated to see if it makes sense in terms of what was originally perceived. The material, the stuff of mental processing, is then projected back into the future for Qualification. Just as correlation was putting things on, Qualification is taking things off. Neither of those concepts is adequate, both are too mechanistic, for complete understanding of this process, but they help to understand the historicity and futurity of the thinking at this level. The qualified material is again cast back to discrimination to be 'tested' against what has been initially perceived to see if it still makes sense, and is then brought forward to Study.

The area of Study has already been entered in the process of qualification and correlation. These phases have been overlaps of Pondering onto Study. The present phase may be considered a final processing or bridge to Representation.

(4) Evaluation—includes representation and experiment. It is the phase in which the conception is placed into a form that can be tried either in formal scientific experiment, or against established systems of knowledge. Experiment

always goes back to the state of pondering for comparison with what could have been, and possibly should be included or excluded (correlation and qualification). Only when evaluation is satisfactory can Realization be claimed.

(5) Realization – includes knowing, estimation, and demonstration. Just as Representation was a kind of tentative knowledge to be evaluated, Knowledge is a satisfied kind of representation to be demonstrated. Estimation is only the fore-cast of imagination in a tentative kind of way so that what could be is included at the stage of pondering for fuller conception.

The diagram on the following page represents the stages discussed. Volition and Emotion are included to demonstrate not only their comparable range, but to indicate those intellectual functions which are impinging upon the other mental areas as being closer to the emotional or volitional line. The diagram should be conceived as an unrolled cylinder, so that the volitional areas which will be included in future diagrams above the volitional line are actually closer to the emotional line. Memory and Evidence should actually be in a plane above that of the page, directly above the intellectual line – they may draw from either emotion or volition.

VOLITION: Stimulation ---> Action

$$Qualify$$
$$\updownarrow$$
Discriminate $\leftrightarrow$ Correlate
$$\updownarrow$$
Memory $\leftrightarrow$ Compare $\leftrightarrow$ Experiment
$$\updownarrow$$
INTELLECT: Ignorance$\rightarrow$ Awareness$\rightarrow$ Wish$\rightarrow$ Regard$\rightarrow$ Assess$\rightarrow$ Ponder$\rightarrow$ Study$\rightarrow$ Represent$\rightarrow$ Know
$$\updownarrow \qquad\qquad\qquad\qquad\qquad\qquad\qquad\qquad\quad \downarrow$$
Imagine ◄─► Estimate $\qquad\qquad\qquad\qquad$ Demonstrate

EMOTION:Discontent $\qquad$ ---> Satisfaction

EMOTION

EMOTION

There are some general terms which apply to all kinds of emotions:

Emotion – that division of the mind which is comprised of a physical/mental reaction perceived as excitation (which may be pleasurable, painful or both). Differentiated from, but related to Intellect and Will.

Sensitivity – the power of emotion.

Feeling – the act in which emotion engages.

Excitation – the subject of feeling.

Affect/Sensibility – the external and internal objects of feeling.

Affect is the external objective view of feelings (i.e. what another being looks like when demonstrating feelings).

Sensibility is the internal objective view (i.e. what one experiences when feeling).

The emotions can be further categorized as prospective, sympathetic, discriminative, and gratification. These steps are roughly analogous to the intellectual groupings of precursory conditions, consideration, conception, evaluation, and realization. The stimulation of emotions remains a mystery. Nevertheless, it is clear that emotions are related to knowledge and relate themselves to Volition. These relationships are not clearly understood, and the following sections, especially the one on **Integration of Mental Functioning** will attempt to explore possible relationships.

The task was arbitrarily set at the end of the preceding chapter of moving from Discontent to Satisfaction. Any emotion other than Discontent which might fit the state of mind in which one becomes aware of Ignorance will not lead to productive mental functioning. This assumption will be explored in the section on **integrated mental processes**.

Prospective Emotions – include desire, courage, and hope. Desire is the necessary outcome of Discontent if mental function is to progress. Persistent Discontent may be necessary to fuel Desire, but if Discontent does not allow Desire to develop, it overcomes mental functioning and creates a state of chronic depression or anhedonia, which by definition excludes Desire. Whether a mild form of blunted feeling can allow desire to arise and permit normal mental functioning is an open question, but clearly any analgesia of feeling will discourage effective progression of the mental process, and require a form of functioning which is not economical and surely must add to the burden of normal functioning. Once Desire has fulfilled its function, the progression of performance requires that an emotion which charges the sensibilities with the prospect of dealing with information which may require dealing with unpleasantness will occur next, and that emotion is called Courage. Courage requires further exploration later, as it entails interaction with Intellectual and Volitional components to arise, but serves as a bridge between Desire and Hope.

Hope may be defined as Desire with intent or possibility, and its necessity in the progression of mental functioning is self evident.

Sympathetic Emotions – include concern and care. In this stage, which could not arise without Hope, Hope is partially suspended, and partially engaged and directed to a specific object of mental function, which may be defined as Concern. The general term, concern, becomes more specific as the intellect analyzes the data, and may be then called Care.

Discriminative Emotions—include excitation/aggravation and affirmation/ negation. At this stage, during which the idea is being conceived, the emotional responses move from the diffuse to the specific in indicating whether or not the intellectual analysis 'feels' right or not. This widely maligned emotional concomitant of intellectual decision may be referred to as intuition, and is an important part of the most rigorous logic, even when denied. The affirmation/ negation stage is called Certification.

Gratification – includes insight and satisfaction. Insight is the emotional counterpart of knowledge; Satisfaction, the analog of demonstrated or proven ideas.

Repeated references have been made to analogies and concomitance between Intellect and Emotion. These are diagramed on the following page:

INTELLECT

PRECURSORY CONDITIONS CONSIDERATION   CONCEPTION
EVALUATION   REALIZATION

Ignorance   Awareness   Wish   Regard   Assess   Ponder   Study
Represent   Know   Demonstrate

Discontent   Desire   Courage   Hope   Concern   Care   Excitation
Affirmation   Insight   Satisfaction

PROSPECTIVE   SYMPATHETIC   DISCRIMINATIVE
GRATIFICATION

EMOTION

VOLITION

VOLITION

Volition -- that division of the mind which resolves mental uncertainty. Differentiated from, but related to Intellect and Emotion.

Ability – the power of volition.

Will – the act in which volition engages.

Choice – the subject of Will.

Action – the object of Will.

Will, the act of Volition, can be cast back into the past, in which case it is called Habit; the future, Intent; and turned upon itself in the present, Determination.

There seems to be both an intellectual and emotional connotation to Will, as well as a denotation of its own. Part of this confusion arises from the variable indications of the word as a state of being, as an act, and as a power (e.g., volition, willfulness, strong-willed). Most states can also be acts (grammatical change of intransitive to transitive verb), and that movement from state to act is volitional. This lies in the fact that every state of being is part of a process (act-ion) – process can be defined as the change from one state to another; or state as slices of process. A power is needed to keep the process going; that power is volition, and it may be sliced into states of being also, which correspond to the intellectual and emotional states at that stage of their respective processes. Actually the correspondence is as a bridge between the states.

Some of the confusion, further, lies in the use of the word *Will* for *Shall,* which confuses ability with mandate; <u>can</u> with <u>must</u>.

The broad groupings of Will are prospective volition, choice, and action. These phases carry the volitional process from stimulation to action.

Prospective Volition – includes motive, intent, and preparation. Motive is the mental formulation of that which gives the power to move the mental processes. It is the result of stimulation; a kind of 'stimulation with awareness'. Intent is a concretized motive. The state of Preparation is a drawing of intent into the area where means and methods are ordered.

Choice – includes **plan** and **course**. **Planning** is as complex a volitional state as **Pondering** was as an intellectual state. It casts itself forward into **Course** (or imagined course), creating the course as it is also limited by the possibilities. This casting forward is a consideration and choice of Methods, which involves attention to the requirements of the situation, as well as the Business (what is done, 'action'), of the subsequent course. In addition,

Planning casts itself back to **Preparation** in assessing the Means available. That involves the Use (of what is required) and the Adequacy (of the Means for the Business at hand). In a way, it can be said that Preparation and Course are the outer limits of Planning, the range with which Plan plans, the bridges from Intent and to Undertaking.

To understand this aspect of volition, it might be useful to imagine the sequential steps of Preparation, Plan, and Course in terms of cooking. If one prepares to cook, the course (recipe or directions) is already in mind. Planning involves a consideration of the method, i.e., the requirements of the recipe. That step must of necessity include the use to which each of the ingredients will be put (means), and of course the adequacy of the kind of utensil and amount of ingredient. The adequacy, in turn, depends on what will actually be done (Business), and only when this kind of planning is completed, can the recipe actually be followed.

Action – includes undertaking, essay, pursuit, and action. Undertaking is analogous to beginning the recipe. If the planning has been inadequate, that false undertaking is an essay, and must return to the planning stage. Pursuit is the actual thing done in completing the undertaking, and action represents not the generic heading of this phase, but the externalization of the efforts.

These steps are represented graphically on the next page.

E: Discontent → Desire → Courage → Hope → Concern → Care → Excite → Certify → Insight → Satisfaction

Qualify
↕
Discriminate ↔ Correlate
↕
Memory ↔ Compare ↔ Experiment
↕

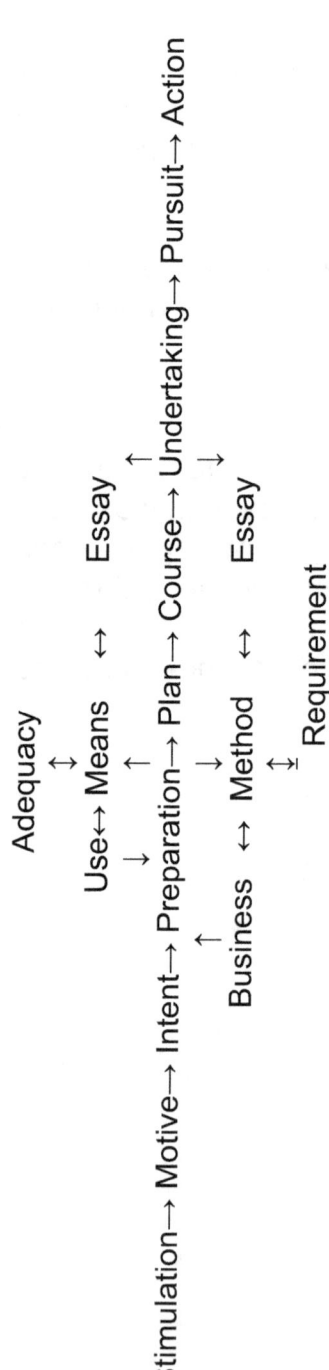

I: Ignorance→ Awareness→ Wish→ Regard→ Assess→ Ponder→ Study→ Represent→ Know → Demonstrate

Imagine ▼—▶ Estimate

Essay
↑
Adequacy
↕
Use↔ Means ↔ Essay
↓
Business ↔ Method ↔ Requirement

V: Stimulation→ Motive→ Intent→ Preparation→ Plan→ Course→ Undertaking→ Pursuit→ Action

INTEGRATION of MENTAL FUNCTIONS

As we examined the three traditional divisions of the mental life (Intellect, Emotion and Volition) which respectively refer to one's reasoning, feeling and choosing, we realize that the mind is not so simply divided. The divisions do interact and influence each other, but not as global behemoths implied in the COMPONENTS OF MENTAL FUNCTIONING (p.25); where we described each division as an independent, perhaps bordering on sovereign, player in the mini-dramas.

We noticed that there was a forward progression within each division. In the division Intellect, the mental processes proceeded from Ignorance -> Demonstration. in the division Emotion , the mental processes proceeded from Discontent -> Satisfaction, in the division Volition, the mental processes proceeded from Stimulation -> Action. These steps are summarized in the table *The entire mental sequence* on page 53.

As we described those steps in the individual chapters on the traditional divisions, there also arose a need for a driving force to allow the mental processes to proceed along the steps. *Ignorance* did not naturally turn to *awareness* just because *evidence* arrived. The driving force was found in the lateral interaction amongst the steps in other divisions at a similar level of mental functioning (*ignorance* in the INTELLECT is made available by *discontent* in the EMOTION to be acted upon by *stimulation* in the WILL to be driven to *awareness,* the next step in INTELLECT. The interaction amongst the first step of each of the three divisions, *ignorance, discontent,* and *stimulation* not only makes more sense than any other interaction, but creates dynamics amongst the other levels and divisions which almost force the creation of analogous stages in comparable divisions.

We will first try to outline that process as it occurs in the best of conditions to show how the functions comprising the steps of the mental processes integrate the processes, and call that chapter **INTERACTION OF MENTAL STATES** to demonstrate how they depend on each other, maintain their integrity and thereby, their authenticity.

Many of our habitual ways of referring to our mental activity are based upon vocabularies which have been extant so long the definitions seem self evident. Indeed, the ways in which we think about ourselves are tied to words which seem to have more meaning than we can easily explain, and we end up saying, "Well everyone knows that."

An example of this phenomenon are the words for **mind** and **soul**, which have been bandied about by centuries of philosophers, and have had their meanings twisted by medical, religious, and psychological proponents. Not only do the meanings get obscured by well meaning investigators, but the words are different in different languages, adding a cultural influence to the meaning.

We have looked at Mental Functioning from the top-down (p.10) by borrowing a technique called thought experiment, pioneered by Galileo and still used by Einstein. This method combined with Husserl's procedure in Existential Phenomenology[23] provided the data from our own introspection of the mental processes ready at hand, and the interview of willing conversationalists in our environment. It is a technique which starts with a phenomenon, an experience, and dwelling in all aspects of it, including one's subjective experience until the meaning of the experience is un-covered in all its essential aspects. By clarifying the meaning of these functions, we will not only see how they fit into a unitary thought experiment, authentically, but also see how they can fit into biologic research by neurophysiologists. Even if they are thought to have long-established meanings, their relationship to the physical aspect, the substrate of mental life, the brain, is not clearly understood; perhaps allowing further progress in bottom-up research, now being done in clinical settings.

On the other hand, a brief diversion into the biology which lies at the base of these systems will be helpful. As we go through an ordinary uneventful day, our senses are awash with information. This information stimulates receptors (eyes, ears, smell, touch, and specialized nerve endings in the interior), are transduced from their natural form (light waves, sound waves, aromas, pressure, etc.) to digitized electronic impulses, which reaches us through extensions of the central nervous system:

1. Cranial nerves – these are direct extensions of the brain, which end mostly in the head, and for our purposes consist of what are usually called the five senses[24]. There is more to the cranial nerves, but this description suffices for our objective.
2. Peripheral nerves – these are extensions of the central nervous system which wander throughout the body, and for our purposes, monitor sensations at the surfaces of the body, which interface with the external world[25].
3. Visceral nerves – these are also extensions of the central nervous system which wander throughout the body, and for our purposes, monitor the internal states of the body[26].

---

[23] Gk. *noumenon* "that which is perceived,"
   phainomenon "that which appears or is seen," "fact, occurrence,"

[24] I – smell, II – sight, VII – taste &touch, VIII – hearing, IX – taste

[25] touch

[26] One remarkable example of this is the visceral afferent system, which reports changes in the internal organs of the body, and compensates for their changes by effecting corrective action through the central nervous system. Control of intestinal peristalsis is one example; endocrine feedback, another. Examples in the Internal Milieu abound, but simple interchanges between the body and the environment also occur frequently, without conscious knowledge.

As we proceed through our day with no changes occurring in the external environment, internal physical conditions, perceptions, or any of the para-physical and metaphysical mental constructs, as we have previously defined them[27]. A steady state exists. There are no changes, or the changes which do occur are familiar, ordinary, and arouse no concern. Everything is copacetic (in the vernacular jargon) or homeostatic (in biological terminology). The use of these two synonyms, *copacetic* and *homeostatic*, goes to the heart of why this book is written. The vocabulary which has grown around the phenomenon of mental functioning is replete with synonyms, many of which have flattened the meaning from which they have sprung, and the complexity to which they refer has not yet been defined.

It puts one in mind of the analogy[28] the Churchlands used in explaining why Albertus Magnus (scholar and philosopher, 1206-1280) failed to explain fire. Ostensibly, the vocabulary of the age did not contain necessary referents to explain the phenomenon adequately. e.g., the element, Oxygen, had not yet been discovered, ergo, fire as an oxidative occurrence could not be adequately explained.

From a neurophysiological point of view, information enters the brain through the channels described, and information is deposited pretty directly into the sensory cortices[29].

The sensory cortices can be thought of as the gateway to consciousness or what we have been referring to as Intellect. Notice, in that statement, the qualifier "pretty directly", which makes quite a bit of difference in how we think of information reaching consciousness. In addition to the input being available to the cortex, or consciousness, it is also intercepted by the thalamic nuclei, which serve as distributors, or way stations, allocating information to selected parts of the brain. The parts of the brain which are selected to receive this information are not chosen consciously, but in a manner which is not completely understood. The result of this dispersal is that not only does the data reach consciousness,

---

[27] *Meaning*, Richard Stevko. The Graven Image, Publishing

[28] Annals of Science:Two Heads:New Yorker, 2/12/2007.

[29] These cortices (cortexes) are relatively small areas on the surface of the brain, which gain direct information as opposed to so called association cortices, which comprise a much larger percentage of the brains surface, and function primarily by interaction with other parts of the brain. The primary cortices are the motor cortex (anterior to the Rolandic fissure), the sensory cortex (posterior to the Rolandic fissure), the visual cortex (in the occipital lobe), the auditory cortex (in the perisylvian area of the temporal and parietal lobes). These are called primary cortices because they have recognizable primary functions as demonstrated by direct stimulation in the operating room on conscious patients, and because of definable losses when these areas are selectively damaged by stoke or injury. They also have microscopic architecture different than other parts of the brain.

The senses of smell and taste are localizable to a great extent, but have more nebulous borders. This seems to be a characteristic of mental functions as they approach what has been called psychological, rather than somatic functions. To be fair, however, somatic functions in the deep viscera are as indefinable as any "psychic" feeling. Witness the difficulty in describing some heart attacks, often confused with indigestion.

where we can act on it, but also it is available to parts of the brain which act on the information in ways which we may not be consciously wise enough to effect, but have been proven to have survival value for our species. The ultimate effect of this system is that not only is information (evidence) available to our Intellects, but act outside the sphere of conscious control, influencing parts of the brain which form the substrate for our Emotional and Volitional aspects of the Mind's activity.

Returning from our biologic diversion to traditional descriptors of mental life, the sensory input which was described above as entering the sensory cortex is perceived by the recipient as registering to the Intellect. Put most simply, the data is processed by going sequentially from the Intellect to Memory, then to Speculation, then to general distribution. That simplification is a deceptive oversimplification, done to provide a structural armature, a framework upon which complex facts can be located. The complex facts to which we allude consist of the many layers and overlays of circuit systems which allows the brain to function with its daunting complexity. The most basic of these systems can be diagramed in the following way:

Events —▶ Intellect —▶ Memory —▶ Speculation —▶ Association Cortices
and Feedback circuits.

A truer picture which reflects some of the complexity and intricacy of brain function and, by extension, the Mind, necessitates the following attachments of known function; and this is still an incomplete picture of mental function. It is wise at this point to be reminded that there are no known structures in the brain that fulfill the requirements of *Intellect, Memory, or Speculation*. These are simply terms we have devised throughout the history of culture and language development to refer to those segments of mental life which they have grown to indicate.

The events which stimulate our sensory receptors and make entry into Intellect as perceptions have already been processed extensively. They have registered on the receptors, and they have been intercepted by the thalamic nuclei and distributed to many association cortices and feedback circuits either before or simultaneously with their registration in the sensory cortices. This seems like a lot of activity for simple perception, but the speed at which neurons carry information in neural circuits is measured in hundredths of a second. At that speed each circuit could register a simple piece of information in at least ten locations each second. The number of circuits is unknown, but estimates have been made that the number of synapses[30] in a single human brain exceeds the

---

[30] connections between neurons.

number of stars in the Milky Way. That's a lot of circuits, and each system has multiple overlays of similar circuits. We have not been created parsimoniously[31].

In addition to simultaneous or imminent registration of data in Intellect, and in association cortices, there is also serial transmission of data from Intellect to the centers which form the substrate for emotion. Of course, the information has probably already been there and in addition to perceiving, as thought, the data from the external world, we also react to it emotionally. It is known that registration of information into Memory (an intellectual function, in our analysis) is aided by emotional content. Indeed, there is a school of thought who holds that Emotion is the fuel that drives the engine of mental life, and that Intellect simply provides the cover story, or allows cognitive content to the emotions for better understanding. Regardless of which comes first, they happen so quickly we cannot discriminate between Intellectual content and Emotional content at the time of occurrence; they happen so quickly, many contend they are two perspectives on the same event.

In addition to the above, the association cortices all have feedback circuits which report back to the originating portion of the brain to acknowledge or deny the veracity or recognition of received information. That feedback is handled as new information and re-reported along with or coloring the new information. In this way, the mind constantly monitors its own balance of information and constantly updates it.

As a result, our oversimplified chart becomes so much more complex that attempts to represent it contain so many arrows coming and going from one site to another that the diagram becomes a tangle of lines. Indeed, the microscopic section of actual brain tissue is such a complex snarl of axons and dendrites that the individual circuits are almost impossible to trace.

For our practical purposes, data which has been received by the Intellect is sent to Memory (1) where the information is compared to previously stored information and is either recognized as familiar (consistent with expectations), or alerted as novel information (unfamiliar data) and alerts the system by returning (2) signals of caution to Intellect, and projects the stored data along with new data from Intellect to Speculation (3). Speculation signals Intellect (4) that no changes are expected, and Memory is given a reinforcing message (5) to store for later retrieval.

Meanwhile, Emotion and Volition respond (6) to the information received by the Intellect from Sensation, as well as that modified by Memory and Speculation. At the same time, signals are sent from Emotion to Memory and Speculation (7), and from Volition to Memory and Speculation (7).

---

[31] one argument for evolution and against intelligent design is that the development of the nervous system, if designed by an Omnipotent Creator, would not have happened so haphazardly.

Their responses to the steady state responses of Memory and Speculation are Reminiscences and Presentiment (Emotion) and Habit and Intent (Volition)[32]. Each response is duly registered by each faculty and stored in Memory for future comparison.

Should the incoming perception change, but still be one that is familiar to Memory and interpreted as benign by Speculation, then the above pattern retains itself in the same manner. Most of our automatic reactions to familiar environments and internal situations are of this sort.

When, however, the incoming perception is not familiar, the reaction is quite different. In the cognitive domain, Intellect, which has been shown something it didn't know, has its Ignorance uncovered. Although the Intellect may not be ignorant in the sense that it lacks information, it acts as though it were, simply because that information with which it is familiar is challenged by the new information which it does not recognize. This new information may not be challenging in terms of requiring a new way of understanding the environment, but it does not have soothing feedback from memory which assures the organism that this information is benign. Any new information, which is different from the last batch processed by the senses, requires acknowledgement by the nervous system as new and must be pigeonholed into familiar slots where it can be stored as reassuring information for comparison with all future data.

As Information increases in the cognitive domain, the ratio of Knowledge to Ignorance increases and in the realm of feeling, Emotion responds with Discontent, that emotion rising proportionally to the Knowledge: Ignorance ratio; and Volition is called upon to settle the uncertainty. It is at this point that the preliminary remarks regarding Intellect, Emotion, and Volition in the preceding sections[33] need to be integrated.

Intellectual Ignorance acts on Emotional Discontent to arouse Volitional Stimulation.

Intellect	Volition	Emotion
Ignorance		Discontent
	Stimulation	
Awareness		Desire

Stimulation, in turn, pressures the Intellect and the Emotion, inducing them to allow Ignorance to progress to something else, and to allow Discontent to change into another emotional parameter. In authentic mental functioning the most productive result occurs when Ignorance is allowed to become Awareness (of the

---

[32] Note that these responses are a casting back (Reminiscence, Habit) and forward (Presentiment, Intent) of Emotion and Volition in their functional evaluation of where the new information fits.

[33] Components of Mental Functioning, p.25

evidence which has manifest ignorance). Stimulation induces Discontent to develop into Desire (to relieve the discontent and ignorance, which are after all only two sides – emotional and intellectual – of the same situation). This complex interaction is an archetype of each step in the mental process. Ignorance and Discontent have both been simulated not only by external factors, but by reinforcement of each other. They are unified and unrelieved; grow together in intensity to incogitancy with anxiety. Each apart and both together, just as both sides of a coin placed in a pocket, act on volition, shaping it, stimulating it, and in turn being shaped and changed, moved, by it.

Neither Ignorance (Intellect's response to novel information [Evidence]) nor Discontent (Emotion's response to novel information [Evidence]) can create Stimulation (Volition's response to the effect Evidence has on both Intellect and Emotion) by themselves. They must act together. An intellect which is ignorant of anything will remain so unless the mind is also discontented; and vice versa (an emotional state which is contented will remain so until the Intellect becomes aroused to its ignorance). In the same way, Stimulation cannot mobilize Ignorance to awareness without Discontent being not only present, but mutually converted to Desire. All other alternatives will be demonstrated in the section on **Uninterested Mental Function**.

The necessity of this tripartite interaction becomes obvious if any two of the states are considered without the third. If, for example, the Volition is not enlisted, or held back for any reason, then Ignorance and Discontent must interact alone. They stimulate each other to Incogitance and Anxiety, respectively, as has been mentioned. Stimulated Discontent without Awareness becomes Anger or Rage. The stimulation in this case must have a different character than the one which grows out of the combination of Ignorance and Discontent (which is a natural form of willingness), and would almost have to be an external, involuntary stimulation. That same kind of external, non-organic stimulation applied to Ignorance, without Desire, results in Stubbornness.

	Intellect	Emotion	Volition
a	Ignorance	Discontent	Stimulated
b	Incogitance	Anxiety	Unstimulated
c	Stubbornness	Anger	Unstimulated

It is of interest that each of the triple groupings:
a)Ignorance, Discontent, Stimulation (i.e., the internal stimulation of Volition) b) incogitance, anxiety, unstimulated (unwillingness, external stimulation) c) stubbornness, anger, unstimulated (unwillingness, external stimulation) are composed of analogous states. Only in the first grouping is there a natural, internal movement to the next stage. The latter two groupings are dead ends without external change influencing them.

To continue the progression of normal mental functioning, Awareness and Desire, each shaping and being shaped by the other, co act, and coupled generate Motive (an awareness of desire, after all; or a desire of awareness). Motive in turn influences Desire, which, without Awareness, becomes Passion; with the modulation of Awareness is Courage. It can be said that courage is an aware desire, stimulated by Motive. Wish can be defined the same way. The difference between the two is not only that one is an emotion and the other an intellect, but that each is the same with a different emphasis.

On the one hand, the mind-brain dichotomy needs to be clearly understood. Platonic and Cartesian systems, which are so deeply ingrained in our assumptions that they seem to be part of the fabric of reality, do not find neat support in recent neurophysiology research on consciousness. These descriptions are far from fully explained, but do raise disturbing questions. As these frontiers become more completely understood, we can expect clearer descriptions, which will subsequently be subjected to the test of time and reality testing. The following issues will be proposed in light of what is so far known, and may speculate on what seems probable, with appropriate caveats. The entire mental sequence can be represented:

A. In tabular form:

INTELLECT	VOLITION	EMOTION
Ignorance		Discontent
	Stimulation	
Awareness		Desire
	Motive	
Wish		Courage
	Intent	
Regard		Hope
	Preparation	
Assessment		Concern
	Plan	
Ponder		Care
	Course	
Study		Excitation
	Undertaking	
Representation		Certification
	Pursuit	
Knowledge		Insight
	Action	
Demonstration		Satisfaction

or B. In mathematical form (with annotations):

1a. Ignorance + Discontent = Stimulation
  When the Intellectual realm is made aware of its Ignorance, and
    the Emotional realm is made aware of its Discontent, then
    the Volitional realm is Stimulated to resolve the imbalance.
1b. Stimulated Ignorance / Discontent = Awareness
1c. Stimulated Discontent / Ignorance = Desire
  As a result of 1a, the Stimulated state of the Volitional realm induces
    the Intellectual realm to convert Ignorance to Awareness , and
    the Emotional realm to convert Discontent to Desire
  as long as neither resists the natural inclination of Volition.

2a. Awareness + Desire = Motive
  When the Intellectual realm is converted to Awareness, and
    The Emotional realm is converted to Desire, then
    The Volitional realm is Motivated to optimize these new states.
2b. Motivated Awareness / Desire =Wish
2c. Motivated Desire / Awareness =Courage
  The motivated state of Volition induces
    Awareness in the Intellect to convert to Wish, and
    Desire in the Emotional realm expresses itself as Courage.

3a. Wish + Courage = Intent
  When the Intellectual realm is converted to Wish, and
    The Emotional realm is converted to Courage, then
    The Volitional realm develops Intent, acting on the other two realms
3b. Intended Wish / Courage = Regard
3c. Intended Courage / Wish = Hope
  The Intent of Volition induces
    Wish in the Intellectual realm to express itself as Regard, and
    Courage in the Emotional realm advances to Hope.

4a. Regard + Hope = Preparation
  When the Intellectual realm is converted to Regard, and
    The Emotional realm expresses Hope
    The Volitional realm develops Preparedness, causing
4b. Prepared Hope / Regard = Concern
4c. Prepared Regard/ Hope = Assessment
  The Preparedness of Volition causes
    The Intellectual realm to show Assessment in lieu of Regard,
    and The Emotional realm to develop Concern.

5a. Assessment + Concern = Plan
 When the Intellectual realm is converted to Assessment, and
  The Emotional realm is filled with Concern, then
  The Volitional realm begins creating a Plan
5b. Planned Assessment / Concern =Ponder
5c. Planned Concern / Assessment = Care
 The Planning of Volition causes
  The Intellectual realm to Ponder, and
  The Emotional realm to express Care.

6a. Ponder + Care = Course (direction)
 When the Intellectual realm is converted to Ponder, and
  The Emotional realm is converted to Care, then
  The Volitional realm develops a Course (direction)
6b. Directed Pondering / Excitation = Study
6c. Directed Care / Ponder = Excitation
 The Direction of Volition causes
  The Intellect realm to Study, and
  The Emotional realm experiences Excitation.

7a. Study + Excitation = Undertaking
 When the Intellectual realm is converted to Study, and
  The Emotional realm is converted to Excitation, then
  The Volitional realm develops an Undertaking.
7b. Undertaken Study / Excitation = Representation
7c. Undertaken Excitation / Study = Certification
 The Undertaking of Volition causes
  The Intellectual realm to create Representation, and
  The Emotional realm to exercise Certification.

8a. Representation + Certification = Pursuit
 When the Intellectual realm is converted to Representation, and
  The Emotional realm is converted to Certification, then
  The Volitional realm develops a Pursuit.
8b. Pursued Representation / Certification=Knowledge
8c. Pursued Certification / Representation = Insight
 The Pursuit of Volition causes
  The Intellectual realm to create Knowledge, and
  The Emotional realm to exercise Insight.

9a. Knowledge + Insight =Action
 When the Intellectual realm is converted to Knowledge, and
  The Emotional realm is converted to Insight, then

The Volitional realm develops Action.
9b. Active Knowledge/ Insight = Demonstration
9c. Active Insight/ Knowledge = Satisfaction
The Action of Volition causes
The Intellectual realm to exercise Demonstration, and
The Emotional realm to exercise Satisfaction

AUTHENTIC MENTAL FUNCTIONING, Overview

INTELLECT

PRECURSORY CONSIDERATION   CONCEPTION   EVALUATION   REALIZATION

Ignorance   Awareness   Wish   Regard   Assess   Ponder   Study   Represent   Know/
Demonstrate

EMOTION

PROSPECTIVE   SYMPATHETIC   DISCRIMINATIVE   GRATIFICATION

Discontent Desire   Courage Hope Concern Care   Excitation Affirmation   Insight Satisfaction

----

VOLITION

PROSPECTIVE   PREPARATION   CHOICE   ACTION

motive intent   preparation plan course   undertaking essay pursuit action

INTERACTION OF MENTAL STATES

Interaction of the Intellectual, Emotional, and Volitional processes begins when evidence impinges on that part of the Intellect to which it is strange, Ignorance. The state of Intellect which is produced, Awareness, has two parts; one objective (awareness of evidence) and one subjective (awareness of ignorance). The usual result of this event is a two part emotional reaction, Alertness (objective) and Discontent (subjective).

It is possible that Discontent will not be produced. The failure of this response may be due to extremely subtle evidence or weak perception of the Intellect. These first two factors are related, and might better be referred to as the ratio of strength of stimulation to strength of reception. Reasons why this might happen include:

- People of low intelligence may not be stimulated by awareness of ignorance – the so-called "ignorance is bliss syndrome"
- In some highly intelligent people; subtle stimuli which escape most people constantly produce an awareness of ignorance resulting in discontent.
- Another reason discontent may not result is a high level of satisfaction. In this case, there is no need to know, or no interest in knowing, and the awareness of ignorance is ignored.
- A further reason may be a judgment that the perceived evidence is unimportant, of no consequence, and not worth the discontent.
- Further, decreased sensitivity of the emotion may make discontent difficult to stimulate.
- High sensitivity may produce so much discontent, that the discreet discontent from this one parcel of evidence is not discriminated from the general discontent already felt.

Whatever the preexisting condition, it is obvious that Discontent acts as a modulating mechanism to determine the influence or strength or even possibility of Awareness entering the mental process. Discontent, created and directed by Awareness (of ignorance) births volitional stimulation, Willingness (to learn.)

Stimulation usually acts on Awareness and Discontent, not individually, but one reaction influenced by the other, to draw them into Wish (to know) and Desire (to know), respectively. If Stimulation acted on Awareness alone, a stimulated Awareness (general arousal) would result. That is a nonproductive, even disruptive state, unless influenced by discontent, which results in Wish. The same is true of Discontent, which would be unhappiness (stimulated discontent), unless awareness influenced it to turn out as Desire.

Wish and Desire, the intellectual and emotional sides of a single state called curiosity, simultaneously reach back to Stimulation (willingness) and cast it forward to Motive. There is a natural break in the intrinsic mental process at this point. Motive is a backward look at the accretion of steps to this point, just as intent is a forward look from the same point. Before considering the events that turn Motive around into Intent, it would be worthwhile to review the process to this point with the intent of examining the parameters and workings of the event which have transpired.

The first level intellectual state, Awareness, both gathered the evidence, and informed the mind of it. The resultant first level emotional state, Discontent, modulated the degree of Awareness entering the mind, and energized the Awareness/Discontent unity to create the first level volitional state, Stimulation (Willingness). Stimulation reached back and activated Awareness/Discontent and projected it forward, that part of the function being characterizable as direction.

The second level intellectual state, Wish, informed Stimulation which provided the setting for Motive (informed wish). That movement was energized by Desire, the second level emotional state, and modulated by it. The modulation referred to here is the dependence of stimulation on the ground flanked by Desire vs. Repulsion. Should that second level emotional state been closer to Repulsion, it is unlikely that Stimulation would become active, but more probably would have activated a flight reaction.

On the basis of this analysis, it is apparent that:

Intellect – gathers (evidence) and formulates (the information into concepts).
Emotion – energizes the system, driving towards solution.
Volition – activates and directs the other two divisions of the mind.

The Intellect reaches outside the mind by its connections to the sensory organs, making it the only demonstrative one of the mental functions which is responsive to external evidence[34] as well as internal evidence, such as Emotion, Will (esp. conscience) and physiologic change.; that seems to restrict Emotion and Will to the inside of the mind.

The first listed functions – gathering, energizing, and activating – are a reaching back in time

The second listed functions -- formulating, driving, and directing -- are a thrusting forward in time.

---

[34] A possible exception to this may be that one can demonstrate emotional reactions to external stimulation, but it seems that the sensory input is still through the receptors associated with Intellectual function. The fact that the information attributed to Intellectual participation may not be at a conscious level. This issue is not clearly resolved. It has become apparent that independent recent work by Pert, Goleman, and Solomon has repeatedly brought into question the interpersonal influence of Emotion. Likewise, Volition has repeatedly been seen as a interpersonal phenomenon in different ways by philosophers Schopenhauer, Nietzsche, Rousseau, and Freud.

In addition to the past and future aspects of each state, there are two kinds of lateral motion in each state. The first is restricted to the state itself. An example is the possibility that Awareness may produce anything from Discontent to Satisfaction. That is the range of possibility. Actually, depending on any of the factors mentioned in the discussion on Discontent, a narrow passage is allowed for the process moving from Awareness to Stimulation. That passage is the lateral motion which modulates. The second kind of lateral motion is between states. The conjunction of the discreet states, Awareness and Discontent, is a lateral motion of those states to produce Stimulation. That lateral motions may be characterized as:

	INTRASTATE	INTERSTATE
Intellect	selection	informing
Emotion	modulation	emphasis
Volition	channeling	decision

Specifically:

Awareness:
- gathers the evidence that impinges on Ignorance
- informs Emotion to stimulate Discontent
- selects, through the dynamic interaction of Awareness/Unawareness, the evidence used in informing, and formulating the evidence into 'that which is not known'.
- formulates Discontent's influence on Volition to produce Stimulation.

Discontent:
- energizes Awareness
- drives Awareness to Wish
- modulates the influence of Awareness through the dynamic interaction of Discontent/Satisfaction, and
- emphasizes Awareness' influence on Volition to produce Stimulation.

Stimulation:
- activates Awareness and Discontent
- directs Awareness to Wish and directs Discontent to Desire (Stimulated Discontent without activated Awareness produces Unhappiness; Stimulated Awareness without activated Discontent produces Anxiety).
- channels Awareness and Discontent to Wish and Desire through the dynamic interaction of Stimulation/Unwillingness., and
- decides that movement will occur by the activation of interaction of Wish/ Desire to form Motive.

At the second level:

Wish:
- gathers information provided by Awareness
- formulates it into a specific idea

- channels Courage in Courage's activation of Motive, and
- selects from the evidence what is to be further formulated.

Desire:
- energizes Stimulation to move to Motive
- drives that process forward
- modulates Stimulation and Wish, and
- emphasizes Wish and Stimulation in the movement to Motive.

Motive:
- activates Wish and Desire
- directs them to Intent
- channels Wish and Desire, and
- decides that movement will occur.

Review of
INTERACTION OF MENTAL STATES
with
NARRATIVE DESCRIPTION OF Brain Interactions

As we go through the day, our sensory systems are constantly barraged with information. As this information reaches the central nervous system, it is processed. Part of this processing is editing. All the editing mechanisms are not known, and the ones we do know about are incompletely understood. Two of the mechanisms we understand imperfectly are:

1. The attention mechanism – human attention is selective. If we pay attention to too much, we become distracted; if we pay attention to too little, we are unfocused[35]. It is also thought that the ability to selectively ignore nonessential information is a sign of intelligence. The selection of which things deserve attention is also determined by our level of fatigue, the energy level under which we function provides our "edge", as our attention advantage is often called. In addition, our level of preoccupation with other matters (our "emotional state"), also distracts from the attention level. In addition, a poorly understood portion of the brain, called the Reticular Activating System, is thought to influence the level of attention. This is usually associated with non-specific arousal, and may be related to the biologic basis of the emotional system.

2. The thalamic nuclei select and redirect information to selected parts of the brain as that information is transmitted from the sense organs to the cortices. This extremely complex area of the brain is far from understood, and may be more subtle than our measuring instruments can be calibrated.

Although our understanding of the biologic basis of mental life is imperfect, our insight into mental functioning continues to be explored with a vocabulary which is poorly coordinated with the language of biology. Someday, we may have correlates of the two languages, but for now we must continue to discuss mental life using the terms Intellect, Emotion, and Volition. We may draw parallels when we can, but must continuously update our understanding of concomitances.

If we return to our hypothetical nervous system proceeding through the day, awash in information, and note only what happens in the Intellect, we find that there are things we do not know and things we do know. The things we do not know, we may call Ignorance; the things we do know; Awareness. The information flooding into the sensory system, we may call evidence. If these designations, Ignorance and Awareness were simply passive receptacles or

---

[35] The Neonatal Behavioral Assessment Scale (NBAS), which was devised by the Harvard pediatrician T. Berry Brazelton, allows for this factor.

mental cisterns, and the flood of evidence simply needed to fill them, we could create a simple hydraulic model in which evidence flows in, Ignorance is decreased and Awareness is increased.

Biology is seldom so neat. What is not known (Ignorance) and what is known (Awareness) is in a steady state and the Emotional realm is in a state of Contentment. When evidence upsets the homeostasis which existed in the mind, the imbalance informs the Emotional realm of the mind and Satisfaction turns to Discontent. The new evidence is seen as not fitting into the existing paradigm of mental balance. The Contentment axis is tipped and that energizes the Knowledge axis to gather and select more information (evidence), thus turning Ignorance to Awareness.

In the Emotional realm, Discontent is energizing the transformation of information into Awareness. As the development of Awareness proceeds, Discontent is quelled and Awareness stabilized. As Awareness is increased in the Intellect, Contentment is also increased in the Emotional realm, there is an activation of both realms, and the heightened level of Awareness, along with the well-being of Contentment, causes a level of Stimulation in the system, which enhances the transformation of the stimulated state.

The transformed stimulated state is paralleled in the Emotional sphere by the transformation of Satisfaction to Desire, which is matched in the Intellectual sphere by the conversion of stimulated Awareness into Wish. This pair Desire (Emotion) and Wish (Intellect) is driving the Stimulated state into a transformation of the Volitional sphere from Stimulation to Motive.

RIGIDIFICATION. Interestingly, if Discontent is fed by another process (like stress) and not allowed to reach quiescence, the strength of Discontent can modulate another Emotional axis (Stimulation ↔ Unwillingness). In normal mental functioning, this axis usually tips to Stimulation as the source of stress is resolved and Discontent is quieted which, in turn activates the further seeking of Awareness. If Stimulation is tipped to Unwillingness, Discontent reigns and both Awareness and Desire are lulled.

AXES OF FUNCTIONS. The three realms of mental life, Intellect, Emotion, and Volition seem to act simultaneously and in concert with each other. It seems simultaneous when it happens, but when we analyze the simultaneity, the analysis slows down the process, and makes it seem cumbersome. In nature the interactions are so fast that what happens in an apparent instant takes many pages to describe.

First, considering the Intellectual realm, in a steady state, the things we do not know (Ignorance) are converted to things we know (Awareness) by contact with evidence. The axis of Ignorance ↔ Awareness is always in a flux. As we live our lives, new evidence constantly arises and is processed by the Knowledge Axis (an Intellectual function) continuously gathering and selecting evidence to turn Ignorance into Awareness.

The bits of evidence that we continually collect are seldom earth shattering, or even recognizable as a discreet piece of information, but the continual barrage of information of which the evidence is comprised are almost continuously being selected by the Knowledge Axis (Ignorance ↔ Awareness) or being denied access as suitable for building Knowledge.

Secondly, in considering the Emotional realm, the Contentment Axis, (Discontent ↔ Satisfaction) is activated. This may be simultaneously with the Intellectual realm, or as a result (depending on whether or not Intellect and Emotion are two sides of the same coin). Sometimes the arousal of Awareness causes Discontent, and sometimes Satisfaction, depending on other factors influencing the atmosphere.

The Intellectual Axis informs the Emotional Axis, and the Emotional energizes the Intellectual. While the Knowledge Axis is busy gathering and selecting evidence, the Contentment Axis is modulating Discontent ↔ Satisfaction, thus activating another Emotional axis (Stimulation ↔ Unwillingness). The Unwillingness arises from the possibility that the development of Awareness may cause greater Discontent, and the Stimulation arises from the possibility that greater Awareness will cause Satisfaction. This subplot in the Emotional sphere is more that a reaction to the development of Knowledge, it is an integral part of how the Knowledge Axis selects evidence in increasing Awareness.

Should the increase in Awareness cause greater Discontent, for any reason, the Stimulation may tip to Unwillingness, feeding back to the selection process in the gathering of evidence, thereby influencing what is kept as Awareness is created.

EVIDENCE STIMULATES. When Unawareness is confronted by evidence, it is being nudged to Awareness. This happens in the Intellectual sphere. At the same time, in the Emotional sphere, Satisfaction is aroused to Discontent. It is not clear whether Unawareness/Awareness and Satisfaction/Discontent are two sides of the same coin, different perspectives on one interaction, or if Satisfaction becomes Discontent as a result of the dawning of Awareness.

Nevertheless, as Awareness grows, gathers and selects evidence, it informs the Satisfaction ↔ Discontent axis, modulating its growth. When Unawareness is awakened, the Emotional parameter responds with Discontent, fueling the growth of Awareness. As Awareness increases, Discontent, energized with new useful information tends more to Satisfaction.

Activities (in parentheses) of various axes (double arrow) which contribute to mental function's processing of Evidence in the first level

Processing of Evidence

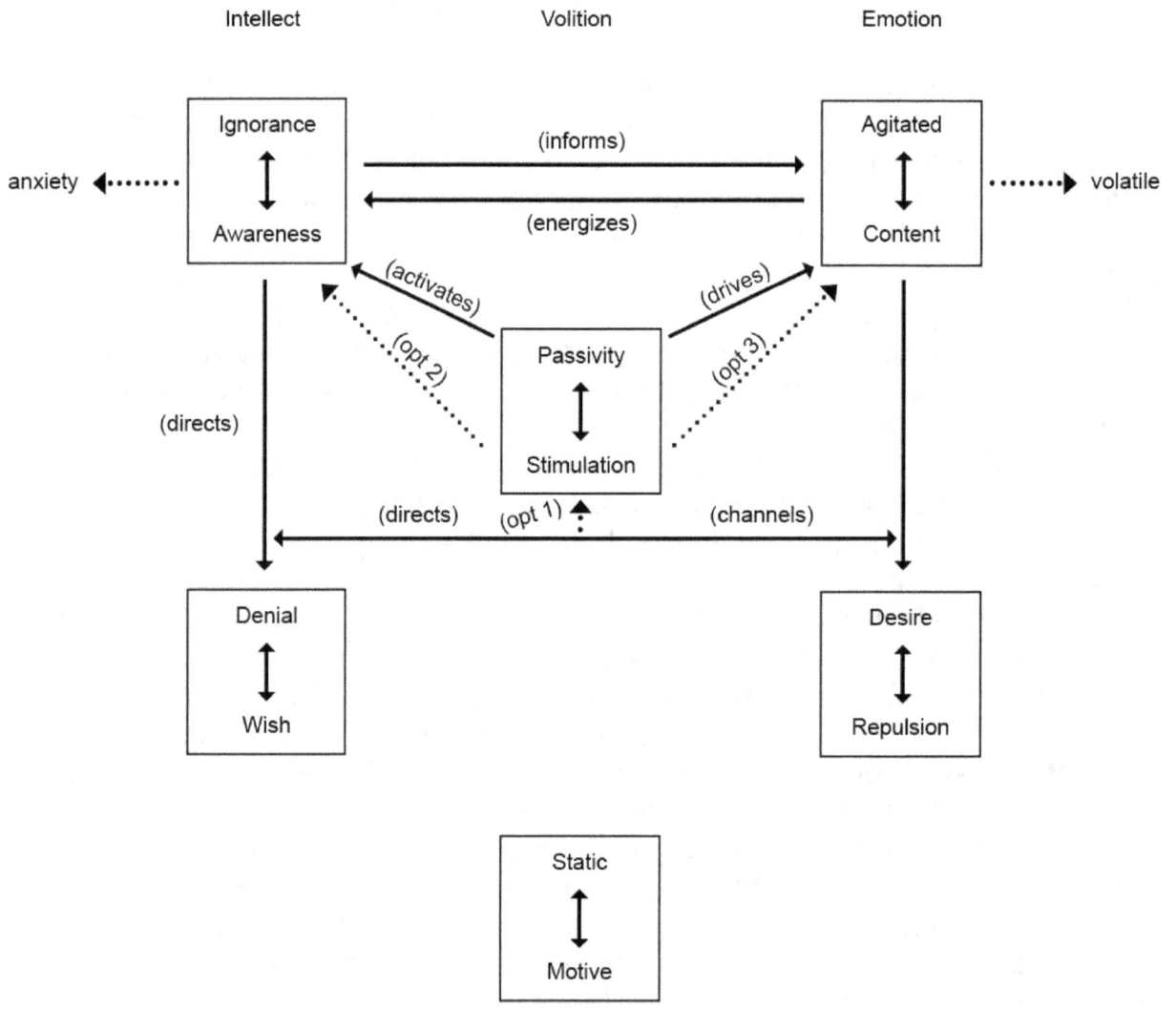

At the second level:

· Wish:
1. gathers information provided by Awareness, and the direction given by Stimulated Discontent,
2. informs Desire and Motive,
3. selects from Awareness specific information (in interaction with Denial),
4. formulates the influence of Desire on Motive.

· Desire:
1. energizes Stimulated Wish,
2. drives Wish and Motive,
3. modulates Discontent,
4. emphasizes Wish's influence on Motive.

· Motive:
1. activates Wish and Desire,
2. directs Wish to Regard and Desire to Hope through activated Desire and Wish, respectively,
3. channels Wish and Desire to Regard and Hope.
4. decides that the interaction of Regard and Hope will move on to Intent.

The remainder of the process can be represented simply, remembering that the interaction of each state is similar to those outlined:

Awareness-----------------------------------v-------------------------------Discontent
Stimulation
Wish ---v-----------------------------------Desire
Motive
Regard -------------------------------------v----------------------------------- Hope
Intent
Assessment -------------------------------v-----------------------------------Concern
Plan
Ponder -------------------------------------v----------------------------------- Care
Course
Study ---v-------------------------------- Excitation
Undertaking
Represent ----------------------------------v-------------------------------- Affirmation
Pursuit
Know --v---------------------------------- Insight
Action
Demonstrate ------------------------------v---------------------------- Satisfaction
Completion

With the aid of the diagram, the discussion can be resumed at the level of Motive, which was defined as a backward look, or resume of all that went before, as Intent is a forward look. The thing which allows motive to become intent is courage. This emotion depends on a) prior success, b) ability, c) desire, e) perceived outcome; in other words, the Emotional concomitant of projected Volition and projected Intellect.

As Motive becomes Intent, Desire becomes Hope (directed by Wish), and Wish becomes Regard (directed by Hope). Regard isw the projection of Wish to Assessment. As Hope grows, Intent is projected into possible Plans for learning. These increases (growing Hope and projected Intent) drive Regard to Assessment. That is the beginning of learning, Preparation. Simultaneously diffuse Hope is consolidated into Concern. Concern is an ambivalent feeling; appropriated, since Assessment is tentative in nature:

Assessment   Preparation   Concern

The future plan is constantly reformulated as Assessment evaluates means and methods and the plan finally consolidated (tentatively),

Plan

And the assessed material ordered:

Pondering   the ambivalence of Concern becomes focused:

Care

And the pondering formulates Plan into:

Course

The course being laid, Pondering becomes study (valuation of ordering data), and Care becomes Excitement. These are concomitants of Course and bridges to Undertaking of the idea formation:

Representation (tentative knowledge)

Affirmation ↔ Negation
(value judgment)

Pursuit (the process of representation and affirmation ↔ negation), which is repeated and repeated until:

Knowledge--- Insight

End of the passive process, if anyone is capable of keeping knowledge and insight from changing their lives:

Action

Demonstration
Satisfaction

Completion

Each step in the process is the fruit of past steps, and at the same time, the seed of the future steps. Each stage draws from the past continually, and returns to the prior steps if he tentativeness of the formulations bear no further fruit. Each stage continually draws from the future, speculatively, for guidance, and progresses to the future stages if each step has been authentic.

What happens when conditions are not functioning optimally have been described somewhat in the chapter **Interaction of Mental States** and will be more closely investigated in **Inauthentic Mental Functioning**. Hopefully, this will serve to demonstrate morbidity and psychopathology in mental functioning.

INAUTHENTIC MENTAL FUNCTIONING

GLOBAL. A failure to integrate the three divisions of the mind consists of generalized dis-integration[36], and considers Intellect, Emotion, and Volition as totalities. All three are involved in Global Inauthentic Mental Functioning. The extreme reaction to a highly malignant stimulus would include Incogitance (Intellect), Numbness (Emotion), and Immobillity (Volition).

The most severe reaction due to failure of integration, total disintegration, or dissociation would be to "die of [37] fright". This unusual phenomenon is usually associated with people who already have compromised cardiovascular systems and involve excess overreaction of the Autonomic Nervous System; specifically the "fight or flight" reaction. A less extreme, but nonetheless debilitating, reaction is a form of the Post Traumatic Stress Disorder of a type which is due to chronic stress.

Other examples of Global Disintegration or Dissociation[38] include:

CATATONIA -- a state of motor immobility, and behavioral abnormality manifested by stupor

DEPERSONALIZATION -- a feeling of separation from yourself and your body

DEREALIZATION -- similar to depersonalization, but it is a feeling of being detached from the external world

AMNESIA -- Some people who experience dissociation have periods of amnesia or "losing time."

TRANSIENT. A reaction which is is the phenomenon of being "frozen with fear", described by LeDoux[39], in which the same symptoms occur (Incogitance, Numbness, and Immobility), but is a temporary event. It happens most typically when being surprised with a threat, and being unable to deal with it, one just freezes. Teleologically, this is similar to the mouse suddenly becoming aware of a hawk and suddenly stops. The involuntary reaction has the advantage of

---

[36] dis-integration - a failure of integration; disintegration - crumbling.

[37] Sir Arthur Conan Doyle, *Hound of the Baskervilles.*
George Engel, *Annals of Internal Medicine,* 1971
Japanese fear of #4 - word pronounced same as word for "death"
Western fear of #13 - associated with arrest of Knights Templar. Also the irregular number which disrupts the completeness of the number 12.
Voodoo death
University Of California - San Diego (2001, December 27). *ScienceDaily* "Scared To Death".

[38] American Psychiatric Association. *Diagnostic and statistical manual of mental disorders, 4th ed, text revision.* Washington, DC, Author, 2000.
Holmes EA, Brown, RJ, Mansell, W, Fearon, RP, Hunter, ECM, Frasquilho, F, and Oakley, D.A. "Are There Two Qualitatively Distinct Forms of Dissociation? A Review and Some Clinical Implications." *Clinical Psychology Review* 25:1-23, 2005.

[39] LeDoux J. Brain mechanisms of emotion and emotional learning. Current Opinions in Neurobiology. 1992;2:191–197.

deceiving the hawk's sharp eyesight for motion, but not necessarily for observing stationary prey; this reaction is a sort of short circuit of information in the brain mediated by the amygdala and bypassing the conscious centers.

PARTIAL DISINTEGRATION involving all three divisions would be fainting or withdrawal. In all other forms of partial disintegration, there seems to be a preponderance of one or two mental function over the others:

> Intellect – intellectualization, numbness, immobility;
> Emotion – panic, incogitance, immobility;
> Volition – reflex, incogitance, numbness;
> Intellect and Emotion – mind races in panic, immobility;
> Intellect and Volition – automatic behavior, numbness;
> Emotion and Volition – panicky behavior, incogitance.

The partial forms of integration have a history in the processes previously described. It is time to return to those. The major question at this point is 'why are non-extreme forms of mental function not consistent with the forms of partial integration.

To return to the beginning, evidence impinged on ignorance. What are the possible ranges of reaction? For Intellect, they can be Unawareness or Awareness. The possibility of unawareness exists if the material is too subtle for the degree of intelligence. To be sure, there are many very subtle events which escape the most astute intelligence...Many times these things are perceived by the Emotion in Discontent. In that case Awareness is eventually aroused and the process begins. Again the question of Awareness is aroused. The degree of intelligence required to recognize emotional stimuli is important. Emotion can respond with Discontent or Satisfaction, and Volition with Willingness or Unwillingness. These possibilities can be combined in various degrees in actual experience:

> Awareness, Discontent, Willingness – authentic mental function
> Awareness, Discontent, Unwillingness – preoccupation, anxiety
> Awareness, Satisfaction, Willingness – familiarity
> Unawareness, Discontent, Willingness – anxiety
> Unawareness, Satisfaction, Willingness – comfort
> Unawareness, Discontent, Unwillingness – anxiety
> Awareness, Satisfaction, Unwillingness – stubbornness
> Unawareness, Satisfaction, Unwillingness – close-mindedness

The intellectual response to unfamiliar evidence may be Awareness of Unawareness. The latter can be due too unintelligence, subtlety of data, or imperception (due to either distraction or inability to perceive such as poor eyesight). In any event, Unawareness is a dead end. It remains in the triad of Unawareness, Satisfaction, and Unwillingness (no need to Will). This is the original state of authentic mental functioning when the environment is familiar. Essentially Evidence has not impinged on Ignorance.

The alternate state, Awareness, can be perceived in varying degrees. Intellect may overreact to the evidence, paying attention to the exclusion of all else. This is affirmation of awareness, but far beyond acknowledgement –beyond even the control of the individual. It is a driven form of intellectual behavior (Preoccupation). I may react in a more controlled way (Awareness); in a neutral way, Ignore-ance (not ignorance, but not paying attention; it may react in a negative way (Disavowal), which is inauthentic; and it may react in a strongly negative way, again uncontrolled, driven (negation).

The first of these reactions, Preoccupation, results in anxiety (strong uncontrolled discontent) and determination (strong uncontrolled Willingness). The lack of freedom of the individual, the driven nature of the mental behavior may be due to the strength of the external stimulus, a high threat situation; it may also be due to a preexisting high level of anxiety or determination. It is a common experience to be so upset about something else, or so determined to 'do something' that the evidence at hand is absorbed beyond reason (Preoccupation). The second state, Awareness, leads to Discontent and Willingness. The third, Ignore-ance, may be due to not-willingness or unfeelingness[40]. The fourth, Disavowal, may occur as a form of inauthentic Awareness, "I didn't see it (though I did)". Disavowal can result from prior satisfaction overwhelming the current awareness (that satisfaction is inauthentic to the current evidence, though it is a sincere satisfaction. Drug highs and Pollyanna attitudes fit this category as well). If it is due to prior satisfaction, the result is unwillingness. If it is due to unwillingness the result is inauthentic satisfaction. The final form, Negation, is an overactive, uncontrolled disavowal. "It wasn't there (though I saw it)". This may result from prior discontent, or unwillingness.

The emotional reactions, likewise, range from uncontrolled positive reaction to uncontrolled negative reaction: anger/fear, discontent, unfeeling, satisfaction, and ecstasy. The last two are pretense. Satisfaction and Ecstasy can be authentic, but not in reaction to an acknowledged strange stimulus. The volitional responses range likewise from Determination, through Willingness, Involition, Unwillingness, to Stubbornness.

The stretching of the original terms, Discontent/Satisfaction and Willingness/Unwillingness, to cover the range of possibilities of reaction to Awareness deserves further definition. Satisfaction and Ecstasy have already been qualified as not the Satisfaction one feels from unawareness of ignorance

---

[40] This is actually an unavailability of the Intellect to process the evidence, although it has been noted. This notation actually comprises awareness and is therefore listed here. It can also result in not-'will or not-Feel as well as result from them.

(or awareness of knowledge) or its energized extension, ecstasy. These secondary forms are more in the sense of denial of discontent or fool's happiness. The volitional range may be rephrased as
Will beyond the control of volition (Determination), inability of 'will (Involition), not-willingness (unwillingness) and not-willingness beyond the control of Will (Stubbornness).

      These complexes fit together:
Preoccupation – Anger/Fear – Determination
Awareness –Discontent – Willingness
Ignore-ance – Unfeeling – Inability
Disavowal – Satisfaction – Unwillingness
Negation –Ecstasy – Stubbornness

      The results of the first and fifth complexes, both of which are beyond control, depend more on circumstance rather than the individual. The states are almost interchangeable (converts make the best Catholics, rabid Democrats change impulsively to strong Republicans, etc.…'the flip-flop phenomenon'). If the energy is increased in these complexes, the result is useless over activity or useless inactivity. That will eventually become complex #3. The only way out is a reduction of energy, a removal from the situation so that control can be established. The second complex is authentic and productive; the fourth inauthentic, negating and counterproductive.

      The second level of authentic mental functioning is the states of Regard (Intellect), Concern (Emotion), and Motive (Volition). Regard is the authentic intellectual response to Willingness, which has grown out of Awareness and Alertness. Inauthentically, the intellect may respond with Preoccupation, Overlooking, Disregard, or Avoidance. Those states are analogous to the ones described for Awareness as Affirmation (not acknowledgement), neutrality, Disavowal, and Negation.

      Preoccupation is a driven form of positive regard which excludes all else and effectively immobilizes the mental process. Overlooking, intentional or not, does not allow the data to enter the mental process. Disregard actively excludes the data, and Avoidance does not allow the evidence to even be considered for regard or disregard, although the mind is aware of it. There are analogous state for emotion and volition. In lieu of listing them, the entire mental process will be outlined with the range of possible reaction, listed by number according to the following code:
1. Driven affirmation
2. Authentic acknowledgement
3. Neutrality, suspension
4. Disavowal
5. Negation

1. Driven affirmation

INTELLECT	EMOTION	VOLITION
Engagement	Absorption	Determination
Preoccupation	Worry	Drive
Assertion	Need	Certitude
Scrutiny	Faith	Predetermination
Obsession	Infatuation	Procedure
Assertion	Passion	Compulsion
Reveal	Certainty	Enforcement
Dogma	Credulity	Confrontation
Command	Ecstasy	Necessity

2. Authentic acknowledgement

INTELLECT	EMOTION	VOLITION
Awareness	Alertness	Willingness
Regard	Concern	Motive
Wish	Desire	Intent
Assessment	Hope	Preparation
Pondering	Care	Plan
Study	Excitement	Course
Represent	Certification	Undertaking
Knowledge	Insight	Pursuit
Demonstration	Satisfaction	Action

3. Neutrality, suspension

INTELLECT	EMOTION	VOLITION
Inattention	Insensible	Involition
Overlooking	Unconcern	Motiveless
Nonchalance	Indifference	Chance – absence of cause
Incomprehension	Apathy	Chance – absence of design
Inattention	Uncaring	Unplanning
Negligence	Dispassion	Uncharted
Latency	Acceptance	Immobility
Ignorance	Unacquaintance	Passivity
Unexpressed	Inappetent	Inaction

4. Disavowal

INTELLECT	EMOTION	VOLITION
Unawareness	Dullness	Unwillingness
Disregard	Negligence	Excuse
Protestation	Satiety	Caprice
Ignore	Doubt	Unpreparedness
Dismissal	Omission	Dissuasion
Contradiction	Relief	Diversion
Disguise	Qualify	Interference
Stupidity	Confusion	Avoidance
Concealment	Dissatisfaction	Coercion

5. Negation

INTELLECT	EMOTION	VOLITION
Ignore-ance	Repellence	Stubbornness
Avoidance	Contempt	Scruple
Denial	Repulsion	Impulse
Disconcert	Despair	Procrastination
Evasion	Violation	Obstruction
Repudiation	Insensibility	Deviation
Suppress	Reversal	Constraint
Taboo	Refection	Abandonment
Prohibition	Desperation	Exploitation

Level 1 of each step is driven behavior. It does not lead to the next step unless it is reduced to Level 2. What is it that causes someone to enter this cul-de-sac? Beginning at step one, Engagement may be the result of evidence which is particularly interesting to other aspects of one's intellect, evidence which has a great deal of appeal (emotional), or evidence which is compelling (choicelessness). All three of these possibilities will lead to emotional cooperation, Absorption. Engagement may also be the result of a highly active intellect, Absorption, or Determination.

So there are three characteristics of the object under consideration (objective characteristics), and three characteristics of the subject under consideration (subjective characteristics):

Objective characteristics:
· Interesting
· Appealing
· Compelling
Subjective characteristics:
· Highly active intellect
· Absorption
· Determination

Of the objective characteristics, it does not seem that any one of them alone can produce Engagement, rather than Awareness. Few objects are interesting to everyone. To the subject for whom the object has interest, it must also have appeal, of some sort. The interaction of interest and appeal produce a compelling force, which engages the subject. From there, Absorption grows from appeal, and finally Determination is the state of volition which describes the will of the Engaged and Absorbed subject. The ocean or the Grand Canyon produces this state almost universally. When it does the mind soars with the degree of interest and appeal one responds to. "I could stay here forever." Of course, no one stays there forever. One of several things eventually happens.

The high energy state can be turned around, voluntarily. Determination can turn negative – a shake of the head and a 'return to reality' results in Ignore-ance. To keep from becoming re-engaged, the view must be repelled with Stubbornness. In effect, the mental cul-de-sac had been entered, and backed out of.

The high energy state can be reduced, voluntarily, by a Determination to be aware of some particular aspect of the object. An awareness of the wave patterns of forms, of the uses of the ocean, of the influence on other people are examples of this step, which results in Awareness-Alertness-Willingness. Both of these processes reversal and reduction of Engagement, can happen involuntarily. A large wave can knock you over, or a shark spotted, will turn Absorption to Repellence, and, if one does not need to be there[41], the presence of the ocean is Ignored with Stubbornness. Daydreaming will reduce Engagement to simple Awareness, or even Inattention.

There is no voluntary way to turn Engagement to Inattention or Unawareness. Those are involuntary, and result from the reduction of the high energy state, or lack of intellectual activity regarding the subject.

The subjective characteristics in themselves also cannot individually produce Engagement. They are the other side of the coin of the Objective characteristics, and must accompany them. It is possible that a single objective characteristic can stimulate the interaction, but both sides must be present.

Ignore-ance, the opposite extreme of Awareness from Engagement, can result, as has been shown, from a turning around, voluntarily or involuntarily, of Engagement. It can also result from Awareness which is threatening (Repellence) or contrary to one's constitution (Stubbornness). This way of

---

[41] If one does need to remain there, Engagement will become Preoccupation, the only way out of the cul-de-sac, and exception to the rule, a continuance of the high energy state, which is not productive to authentic mental functioning, but serves well as a guarded state in a threatening situation – a form of anxiety. Usually this is counterproductive, unless it serves to protect while a way out is found.

presenting Ignore-ance, in contrast to that of Engagement consolidates the fact that the subjective and objective characteristics are conjoined.

Inattention, the neutralization of Awareness, can result from prolonged Awareness (decrease of energy, familiarity) as an economical convenience so we don't have to be aware of everything all the time. It can also arise as a primary response to new information when that information is uninteresting (objective) or the intellect is inattentive (subjective. The emotional correlates of this state are Unappealing (objective) and insensible (subjective); the volitional, uncompelling (objective) and involitional (subjective).

Unawareness, the disavowal of Awareness, results from a lack of intellectual activity regarding the subject. The objective correlate is Subtlety. The emotional correlate is unimpressive (objective) and dullness (subjective); the volitional, undemanding (objective) and unwilling (objective).

The interrelationships of the levels at each step are the same as those described for step one. They are outlined as lists in appendix IV, and in tabular form p. 68. At this point the relationship between steps in inauthentic mental functioning will be introduced.

In the section on authentic mental functioning, it has been describe how Awareness and Alertness produce Willingness, which in turn causes Awareness to become Regard; and Alertness, Concern. To be more accurate, the volitional state which one is in when Aware and Alert is called Willingness, and unless that is voluntarily disturbed, Regard and Concern arise to replace the former steps. This process is automatic, and most of the time instantaneous. When it is disrupted, inauthenticity arises. It has been repeatedly stated that the inauthentic forms are dead ends. The exceptions deserve attention now, as the inauthenticity deepens with progression of that process.

Engagement-Absorption-Determination may be a dead end either in the sense of stopping the mental process from proceeding altogether, or as a temporary suspension for more active involvement at that step. The former may reflect a constitutional overactivity of one of the divisions of mental function; the latter, an immersion, such as the stimulation which precedes creative activity as opposed to prolonged study such as occurs later in Pondering. If it is constitutional, it may progress to Preoccupation-Worry-Drive, the analogous high level, inauthentic step next in the mental process. If the energy level is high, the process proceeds at level1 for each step. This may be characterized as a prejudice in favor of the evidence, and has the disadvantage of consuming a larger amount of energy that he evidence deserves and of not allowing the negative aspects of the evidence inter into the process. Needless to say it is one sided.

The same thing may be said of Ignore-ance / Repellance / Stubborness, in the negative direction. It may represent a suspension of Awareness complex to test the negative aspects, or to dissipate energy against the evidence (constitutional negative prejudice). It has the disadvantages of not considering the positive aspects of the evidence, and of consuming too much energy.

Level 4 is either a mild form of level5, or a devil's advocacy. If not a progressive process, it may simply be contrariness, of obstructionism.

Level 3 is either unawareness, suspension from consideration in the interest of nothing other than to consider something else, or a form of activity which is influenced by the data without responding to it.

All of the forms of inauthentic mental functioning are classified as such because they do not allow an authentic response to the evidence. In each case, they bring something to the process which is not intrinsic to the nature of the evidence. This is counterproductive either in terms of an inappropriate conclusion, or the amount of psychic energy necessary to accomplish the task.

concrete / abstract - a distinction naturally naturally made between entities that one could touch (concrete) and those that were ideas. This difference held until after Rene Descartes[42], raised consciousness levels by referring to them as *res extensa* (things extended...into space) and *res cogitans* (things thought).

Although this does not seem to be a very different from the earlier concepts, it provided future ontologists (philosophers who studied the nature of Being) with challenges.Early amongst these was Bernard Bolzano[43]. His work as a mathematician and logician influenced Charles Sanders Pierce[44], who in turn influenced John Dewey[45] and William James[46], thereby influencing the movement of Pragmatism. Bolzano's 1837 work, *Theory of Science* was groundbreaking amongst logicians, and his ideas on the concrete / abstract issue influenced Franz Brentano[47], who in turn taught Kazimierz Twardowski[48] and Edmund Husserl[49] , thereby influencing Analytical Philosophy and Phenomenology, respectively.

Friedrich Ludwig Gottlob Frege[50] agreed with Bolzano that appeals to intuition have no place in logic. He expanded that these by maintaining numbers, and eventually thoughts, are not external concrete things nor mental entities, but exist in a third realm. This thinking has been incorporated by most f the ontologists from Bozano, on; though much of philosophy has incorporate the third realm into the Abstract realm. Gideon Rosen[51] has confronted the idea of mind independence in contrast to Frege.

---

[42] Descartes (1596-1650), French philosopher, mathematician, and writer, who has been dubbed the 'Father of Modern Philosophy'

[43] Bolzano (1781-1848), a Bohemian mathematician, logician, philosopher, theologian, Catholic priest and antimilitarist.

[44] Pierce (1839-1914), a philosopher, logician, mathematician, and scientist

[45] Dewey (1859-1952) an American philosopher, psychologist and educational reformer

[46] James (1842-1910) an American philosopher, psychologist and physician

[47] Brentano (1838-1917), German philosopher and psychologist

[48] Twardowski (1866-1938) Austrian-born Polish Analytical Philosopher

[49] Husserl (1859-1938), founder of Phenomenology

[50] Frege (1848-1925) a German mathematician, logician, and philosopher

[51] Rosen American philosopher

Although Descartes, the instigator of this animosity at one end of the historical ontology[52], may have poured fuel on the fire by using the term *res* (thing) which may have had a subliminal effect of objectifying the abstract. Nonetheless, at the other end of the historic arch, Willard Van Orman Quine[53] is said to have coined the term *abstract object*[54]. Whether or not the term has a more specific meaning in Analytic Philosophy and mathematics, in this work we will avoid the term object when speaking of abstract concepts. It seems that eventually we will need to unify terms between abstract and concrete, but until then will refer to objects in both spheres as *entities*.

APPENDIX II

structure of individual soul	function	analogous body parts	symbolized caste of society
appetite	Emotion productive	abdomen	Workers
spirit	Volition protective	chest	Warriors or Guardians
reason	Intellect governing	head	Rulers or Philosopher King

---

[52] It could be argued that the distinction existed in Plato's distinction between Real and Ideal, but there was not the strong ontological difference that eruted after Descartes. It could as well be defended that Plato saw the Ideal as a cause of the Real

[53] Quine (1908-2000) American philosopher and logician

[54] Armstrong, D.M. (2010). *Sketch for a systematic metaphysics*. Oxford: Oxford University Press. p. 2.

APPENDIX III

No.	Emotion	Intellect	Volitional Resolution
1	affirms	negates	"I will do it and nothing can stop me."
2	ignores	negates	"Nothing can stop me."
3	affirms	disavows	"I will do it; no reason not to."
4	affirms	ignores	"I will do it."
5	affirms	acknowledges	"I will do it, but shouldn't"
6	acknowledges	negates	"I want to, nothing can stop me."
7	acknowledges	disavows	"I want to; no reason why not."
8	acknowledges	ignores	"I want to."
9	ignores	disavows	"No reason not to."
10	ignores	acknowledges	"I shouldn't."
11	disavows	ignores	"I don't want to."
12	disavows	acknowledges	"I shouldn't, and don't want to."
13	negates	acknowledges	"I shouldn't; nothing can make me."
14	acknowledges	affirms	"I want to, but won't."
15	ignores	affirms	"I won't do it."
16	disavows	affirms	"I won't do it and don't want to."
17	negates	ignores	"Nothing can make me do it."
18	negates	affirms	"I won't do it and nothing can make me."
19	negates	negates	"Nothing can make me or stop me."
20	affirms	affirms	"I will do it and I won't do it."
21	acknowledges	acknowledges	"I want to, but shouldn't."
22	disavows	disavows	"I don't want to, but no reason not to."
23	ignores	ignores	"…"

Appendix IV -- Steps in Authentic and Inauthentic Mental Functioning

The entire mental process will be outlined with the range of possible reaction, listed by number according to the following code:
1. Driven affirmation
2. Authentic acknowledgement
3. Neutrality, suspension
4. Disavowal
5. Negation

INTELLECT	EMOTION	VOLITION
1. Engagement	Absorption	Determination
2. Awareness	Alertness	Willingness
3. Inattention	Insensible	Involition
4. Unawareness	Dullness	Unwillingness
5. Ignore-ance	Repellence	Stubbornness
1. Preoccupation	Worry	Drive
2. Regard	Concern	Motive
3. Overlooking	Unconcern	Motiveless
4. Disregard	Negligence	Excuse
5. Avoidance	Contempt	Scruple
1. Assertion	Need	Certitude
2. Wish	Desire	Intent
3. Nonchalance	Indifference	Chance[55]
4. Protestation	Satiety	Caprice
5, Denial	Repulsion	Impulse
1. Scrutiny	Faith	Predetermination
2. Assessment	Hope	Preparation
3. Incomprehension	Apathy	Chance[56]
4. Ignore	Doubt	Unpreparedness
5. Disconcert	Despair	Procrastination

---

[55] Chance (4) – absence of cause

[56] Chance -- absence of design

1. Obsession   Infatuation   Procedure
2. Pondering   Care   Plan
3. Inattention Uncaring   Unplanning
4. Dismissal   Omission   Dissuasion
5. Evasion   Violation   Obstruction

1. Assertion   Passion   Compulsion
2. Study   Excitement   Course
3. Negligence   Dispassion   Uncharted
4. Contradiction Relief   Diversion
5. Repudiation   Insensibility Deviation

1. Reveal   Certainty   Enforcement
2. Represent Certification   Undertaking
3. Latency   Acceptance   Immobility
4. Disguise   Qualify   Interference
5. Suppress   Reversal   Constraint

1. Dogma   Credulity   Confrontation
2. Knowledge Insight   Pursuit
3. Ignorance Unacquaintance Passivity
4. Stupidity Confusion   Avoidance
5. Taboo   Refection   Abandonment

1. Command   Ecstasy   Necessity
2. Demonstration Satisfaction   Action
3. Unexpressed   Inappetent   Inaction
4. Concealment   Dissatisfaction Coercion
5. Prohibition   Desperation   Exploitation

www.ingramcontent.com/pod-product-compliance
Lightning Source LLC
Chambersburg PA
CBHW080818170526
45158CB00009B/2460